World Symposium on Smart Materials and Applications (WSSMA 2020)

IOP Conference Series: Materials Science and Engineering
Volume 958

Harbin, China
17 - 19 July 2020

ISBN: 978-1-7138-3171-6
ISSN: 1757-8981

Printed from e-media with permission by:

Curran Associates, Inc.
57 Morehouse Lane
Red Hook, NY 12571

Some format issues inherent in the e-media version may also appear in this print version.

Additional copies of this publication are available from:

Curran Associates, Inc.
57 Morehouse Lane
Red Hook, NY 12571 USA
Phone: 845-758-0400
Fax: 845-758-2633
Email: curran@proceedings.com
Web: www.proceedings.com

TABLE OF CONTENTS

ANALYSIS OF THE STEEL STRUCTURAL RELIABILITY UNDER THE BIG DATA 1
Fang Yongfeng, Wang Li, TEE Kong Fah

SYNTHESIS AND CHARACTERIZATION OF PROPERTIES OF ZINC COATED THAUMATOCOCCUS DANIELLII EXTRACTS FOR SOLID STATE APPLICATION 6
M.E. Emetere, M. Ahiara Ikechukwu

SOLID STATE DEVICE FROM PLANT EXTRACTS 12
M.E. Emetere, C. Chijuo, S.A. Afolalu

EFFECTS OF THERMAL DIFFUSIVITY ANALYSIS AFTER IRRADIATION 18
Ahli K. D. Willie, Hongtao Zhao, Teplinskaya Svetlana

LEARNABILITY FACTORS FOR INVESTIGATING THE EFFECTIVENESS OF AUGMENTED REALITY SMART GLASSES IN SMART CAMPUS 24
Sayera Hafsa, Mazlina Abdul Majid

APPLICATION OF SMART MATERIALS IN CIVIL ENGINEERING AND ARCHITECTURE 29
SZOLOMICKI Jerzy

APPLICATION OF DATE PALM ASH TO IMPROVE STRENGTH CHARACTERISTICS OF SABKHA SOIL 35
Zaheer Abbas Kazmi

MECHANOCHEMICAL PRODUCTION OF UREA-CITRIC ACID COPOLYMER 40
Eval O. Miranda Jr., Victor H. R. Silva, Mirtânia A. Leão, Elaine C. M. Cabral-Albuquerque, Silvio Cunha, Rosana L. Fialho

EVALUATION OF THE ENERGY DRIVING PERFORMANCE OF A COOLING SYSTEM ASSEMBLED WITH A PELTIER MODULE OPERATED IN HOT CLIMATES AT DIFFERENT ELECTRICAL CURRENTS 45
David J. Jimenez-Rodriguez, Carlos A. Sanchez-Ramos, Diana Palma-Ramirez, Juan Carlos Paredes-Rojas, Abelardo I. Flores-Vela, Facundo J. Marquez-Rocha

Author Index

IOP Conf. Series: Materials Science and Engineering **958** (2020) 011001 doi:10.1088/1757-899X/958/1/011001

Preface

The WSSMA 2020 is the 2020 World Symposium on Smart Materials and Applications (WSSMA 2020) had planned to take place in Bangkok, Thailand, on July 17-19, 2020. But currently, the entire world is struggling against the virulent pandemic COVID-19, so that the WSSMA 2020 conference changed the presentation method from Oral presentation to Remote Video Presentation to help with Global Quarantine Activities.

The Virtual of WSSMA 2020 was held on Incheon National University at Video Conference Meeting Room on July 17-19, 2020. The WSSMA Committee members attended conference at on-site and also authors attended the conference as Video Meeting. The conference program covered Invited Keynote Speech, Oral, and Poster Presentations, they presented their presentation 15 minutes including Q&A time.

This conference will bring together leaders from industry and academia to exchange and share their experiences, present research results, explore collaborations and to spark new ideas, with the aim of developing new projects and exploiting new technology in this field.

The committee of WSSMA 2020 expresses their sincere thanks to all authors for their high-quality research papers and careful presentations. All reviewers are also thanked for their careful comments and advices. Thanks are finally given to IOP Publication as well for producing this volume.

The Organizing Committee of WSSMA 2020

Committee Chair

Prof. Mosbeh Kaloop

Incheon Disaster Prevention Research Center

Incheon National University

IOP Conf. Series: Materials Science and Engineering **958** (2020) 011001 doi:10.1088/1757-899X/958/1/011001

Conference Photograph

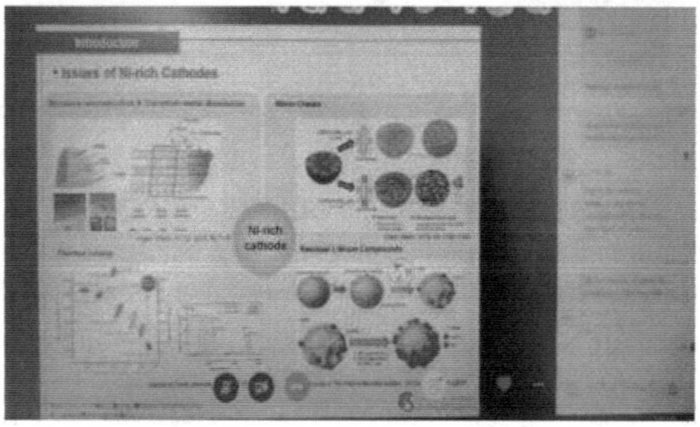

IOP Conf. Series: Materials Science and Engineering **958** (2020) 011001 doi:10.1088/1757-899X/958/1/011001

Sponsors

Incheon Disaster Prevention Research Center (IDPRC)	IDPRC 인천대학교 방재연구센터 Incheon Disaster Prevention Center

Recently, various efforts to prevent and prepare are vitally needed for prevention of Disasters and calamities. So we understand the necessity for technology of disaster and we built up the Incheon Disaster Prevention Research Center (IDPRC) in Incheon National University (INU) in 1997.

Accordingly, Incheon Disaster Prevention Research Center (IDPRC) in Incheon National University has progressed of research on the prevention of disasters and calamities through the various seminars, conference and lectures. These research could be conducted cooperation with Incheon National University (INU) in various fields' structure, soil, hydraulic and environment.

Incheon Disaster Prevention Research Center (IDPRC) will try to be a leader in the disaster of industry through the various research activities and global conference.

IOP Conf. Series: Materials Science and Engineering **958** (2020) 011001 doi:10.1088/1757-899X/958/1/011001

Lists of Committees

Committee name	Organizer name	Organization
Honorary Chair	Dongyan SHI	Harbin Engineering University
Conference General Chairs	Mosbeh Kaloop	Incheon National University
Honorary Chair	Dongkeon Kim	Dong-A University
Technical Committees	J. W. Hu	Incheon National University

IOP Conf. Series: Materials Science and Engineering **958** (2020) 011002 doi:10.1088/1757-899X/958/1/011002

Peer review declaration

All papers published in this volume of IOP Conference Series: Materials Science and Engineering have been peer reviewed through processes administered by the Editors. Reviews were conducted by expert referees to the professional and scientific standards expected of a proceedings journal published by IOP Publishing.

- **Type of peer review:**
  : Triple-blind (Three of Committee Members)

- **Conference submission management system:**
  : Our conference has Secretary for management on conference papers. So authors send their papers to Conference E-mail, and our secretary has managed it.

- **Number of submissions received:**
  : Especially in this time, only few of papers were submitted cause of COVID-19.
  The number of submission papers are 40 papers.

- **Number of submissions sent for review:**
  : The submission papers are 25 papers for review.

- **Number of submissions accepted:**
  : The accepted papers are 12 papers.

- **Acceptance Rate (Number of Submissions Accepted / Number of Submissions Received X 100):** 30 %

- **Average number of reviews per paper:**
  : Three times review on per paper.

- **Total number of reviewers involved:**
  : Three of reviewers are took part in the review process.

- **Any additional info on review process:**
  : We send the Review Materials to author through E-mail, so that author can check the three of review comments papers about their papers. This process also gives a last chance to authors for revision.

- **Contact person for queries:**
  : Jack Son (E-mail: wssma2020@wssma-conf.org)

IOP Conf. Series: Materials Science and Engineering **958** (2020) 012001 doi:10.1088/1757-899X/958/1/012001

Analysis of the Steel Structural Reliability under the Big Data

Fang Yongfeng[1]*, Wang Li[2], Kong Fah TEE[2]

[1]School of Mechanical Engineering, Guizhou University of Science Engineering, Bijie, China

[2]Department of Civil Engineering, University of Greenwich, Kent, ME4 4TB, United Kingdom

E-mail: fangyf_9707@126.com

Abstract: In this paper, the reliability of steel roof under the big data loads is analyzed. Firstly, the roof of steel structure is shown. The relationship between roof span and stress composed of hollow steel joist beam is obtained, and the parameters of the structure of the open web steel joists that is made by the wide flange steel is given. Secondly, according to the random distribution parameters of steel roof and the big data loads, the limit state function of steel roof structural calculation is given. The function is an implicit function. Monte Carlo method is used to calculate reliability index of the steel structural roof with considering the dead loads and the big data loads. Finally, the reasonable suggestions for the reliability and safe operation of the steel structural roof are given from the reliability of the calculated of the steel structural roof.

1. Introduction

In modern manufacturing companies, big data is being captured by using lasers, sensors, wireless networks, etc. These big data include the processing of manufacturing processes, the effects of temperature, vibration, reliability data, etc. The big data is processed in the computer by using different algorithms for improving the products and to increase the competitiveness of the enterprise in the market. Big data is widely used to discover and develop new technologies, methods and decisions. Big data has been paid attention and researched in the modern society. The 5V (volume, velocity, variety, veracity, value) of big data have been defined and they are believed as important as the meaning of new technology [1].

The application principles of big data in the field of security engineering have been analyzed, the theoretical system of the secure big data application was established, the safe big data processing procedure was given, and the theoretical guidance was provided for the application of big data in the field of security science and engineering [2]. With the massive operational data of the equipment, the big data processing technology could be applied to the reliability research of the large-capacity power and electronic systems.

The reliability research has been developed by using the latest theoretical results of big data science based on the data ubiquitous relationship. The big data is done with distributed algorithms, and a feasible algorithm was given by [3].

IOP Conf. Series: Materials Science and Engineering **958** (2020) 012001 doi:10.1088/1757-899X/958/1/012001

The application of big data in traffic reliability operation has been put forward. It is believed that the new opportunities will be brought to innovation in traffic management and decision-making paradigm [4]. The data quality and data characteristics were analyzed, and the highway reliability estimation methods have been proposed [5]. The reliability of the system was predicted and analyzed using big data, and the prediction model was developed. The proposed method has been compared with the existing methods and their prediction model is better the others.

For most engineering structures, massive data would be generated during their service lifetime such as bridges, wind towers, pipelines and automobile chassis. These structures are subjected not only a large amount of imposed load, but also other loads such as wind, corrosion, vibration, etc. The data is acquired timely, and the data would be analyzed and sorted. The structural reliability can be predicted by using the data, and thus the effective maintenance and maintenance measures for the structure are proposed [6].

The existing bridge structures are normally evaluated using big data. It is suggested that the early maintenance of the bridge is very important [7]. A faster detection method which is nonlinear and more practical and feasible has been proposed by using the big data which is collected from the structural health monitoring for a large reinforced concrete structure in Italy [8]. The wind turbine failure model and the maintenance decision system is also established by using the big data which is obtained from the weather and wind power equipment [9]. In most cases, the big data has no obvious distribution rules.

The large-span steel roofs are currently used in a variety of applications in the worldwide such as railway stations, airports, warehouses, etc. The design parameters and the weight of the steel roof was studied, the minimum parameters and precautions for the steel roof design was shown. The reliability of steel structural roof under random loads was studied, the reliability of steel structural roof was changed with time response [10]. The failure mechanism of buildings under snow load was analyzed [11]. It is considered that the multiple failure factors of buildings with the snow load increasing, which increases the components of building failure. For the steel structural roof, during its service period, the steel structure roof not only bears the wind, dead load but also bears the other loads such as wind, snow, etc. The big data has been obtained timely, and the bid data would be analyzed and sorted, the reliability of the steel roof would be accurately analyzed. It is possible to provide strong data and policy support for the operation of these steel structural roof, so as to propose effective maintenance and maintenance measures. The structural reliability from time response has been studied [12], it is considering that the reliability of the structure is time-dependent due to several factors. The damage of buildings under snow load was established from the perspective of physicality and economy, it is considered that it is very important for the early maintenance and maintenance of steel roofs.

In this paper, the big data of the loads is collected from the steel roof beard the loads. The reliability of the steel roof is supported by the open web steel joists is analyzed under considering the uncertainty of load. The reliability of the steel structural roof is obtained by using the Monte Carlo method to simulate the static load and the capacity of the steel structural roof. Finally, it is shown that the method is feasibility and effectiveness by an example.

2. The Steel Structural Roof

The open web steel joists that is pre-made is the main component of the steel roof. It is consisted by two truss beam who are placed horizontally up and down, the up is named the winding, the down is named lower chord, the middle of the winding and the lower chord are place vertical the trusses and the middle trusses named the steel mesh. It is shown as Fig. 1.

Figure 1. The open open web steel joists.

The upper and lower chords of the open web steel joists are composed by the wide flange steel, the wide flange steel is H beam steel. The shape of the steel is that the width of the flange is greater than or equal to the web height. The beam of the general steel structural roof would be made by using the wide flange. The merit of the structure is the section modulus, moment of inertia and strength of the section steel are significantly better than the same heavy others.

The weight of the steel structural roof is made by using the open web steel joists 10 to 40 percent is lighter that other steel structural roof, and its bearing capacity is more than other steel structural roof. Moreover, the wide flange steel has the advantages of wide flange edge, thin web and many specifications, so that it has the advantages of flexible splicing and beautiful appearance when manufacturing steel structural roof. In addition, it is easy to assemble into various components because its outer sides of the flange are parallel, the edge ends are right angles. It is saving the work about 25% when the welding and riveting would be done, which can greatly speed up the construction speed of the project and shorten the construction period.

When designing the steel structural roof, the self-weight and big data load would be fully considered. The steel structural roof is composed by the open web steel joists that has the large span and the greater stress. The relationship between the span of the roof and the average stress of the roof is shown in Table 1.

Table 1. The relationship of the mean of the stress and the span of the roof.

Span（mm）	6000	12000	18000	24000
Stress（MPa）	0.00311	0.00528	0.0072	0.0077

It is shown by Table 1, the span is larger, and the dead load are larger too. It is shown that the big data loads would be studied while the dead load of the steel structural roof must be considered.

The steel structural roof made of the open web steel joists is a square or rectangular grid made up of several horizontal and vertical beams. The steel structural roof is covered with some skins to form a roof that is sheltered from wind, snow and rain. The parameters of the steel structural roof are shown as Table 2.

Table 2. The parameters of the steel structural roof.

Grid size(mm^2)	Grid number	Design strength

520000	6*9	1.03
1000000	14*22	1.32
700000	8*10	1.26
610000	10*12	1.14

3. The Reliability of the Steel Structural Roof under the Big Data Loads

The dead load is noted as *D*, its stress is noted as *d*, the big data loads is noted as *B*, its tress is noted as *b*, the limit loads of the steel structural roof is noted *R*, its strength is noted as *r*. The reliability index of the steel structural roof is noted as β, above various are all random, the limit state function is shown as follow.

$$L=R-D-B \tag{1}$$

If $L>0$, the roof is reliability, or else, the roof failure.

Eq. 1 is an implicit function; it would be computed by using the Mont Carlo method.

The probabilistic distribution of the dead load of the roof and the big data load is shown as Table 3.

Table 3. The distribution parameters of the probabilistic distribution of the dead load of the roof and the big data load.

Loads	Distribution type	Mean (MPa)	Coefficient of correlation
Dead	Normal	0.00869	0.10
Big data loads	Lognormal	0.00445	0.44

The random parameters of the open web steel joist are shown as Table 4.

Table 4. The random parameters of the open web steel joist.

	Distribution type	Mean	Coefficient of correlation
Elastic modulus	Normal	1.7×10^5MPa	0.076
Yield strength	lognormal	240MPa	0.11
Section modulus	Normal	8.0cm	0.05

The Eq. 1 can be computed by using the Mont Carlo method according to the Table 1 to 4. Its reliability index is 1.45, the reliability of the steel structural roof is 0.9265.

It is shown by the result, the reliability of the steel structural roof is high in the service period. Although, the roof must be cleaned the snow and checked out and reinforced the thick part in the extreme weather.

4. Conclusion

The reliability of steel structural roof is computed under big data loads. According to the safety rules of the building and serviceability objectives, the Monte Carlo method is used to consider the uncertainty of roof load and the big data loads on the roof. The reliability of the steel structural roof under the big data loads was obtained. However, it should be noted that this reliability assessment is usually less than 25-45%. Although the reliability is similar, even if the roof design has the same load exceeding 50 years of the load, the substantial difference in probability shape will result in the distribution of the maximum snow load per year. Producing different reliability results, these are important for roof snow loads.

In additional, the application of this analysis method should be fully consider the difference climate and geological distribution. The annual maximum snow load has a great impact on the reliability analysis. At the same time, changing the physical characteristics of the roof has little effect on the reliability index. Secondly, there is a strong correlation between the reliability and maintainability of steel roofs. Finally, for the steel structural roof, the design of the deflection limit results in a larger cross-sectional dimension, resulting in considerable improvement in safety, maintainability and reliability.

References

[1] A Mandawat, A E Williams and S A Francis 2017 Cardio-oncology the role of big data, Heart Fail. Clin. **17** 403-408
[2] Q Ouyang, C Wu and L Huang 2016 Research on basic principles of applications of big data in field of safety science China Safety Sci. J. **26** 11 13-19
[3] M S Al-kahtani and L Karim 2017 An efficient distributed algorithm for big data processing Arabian J Sci Eng **42** 8 3149-3157
[4] E O Ekpiwhre, K F Tee, S A Aghagba and K Bishop 2016 Risk-based inspection on highway assets with category 2 defects Int J Safety Security Eng. **6** 2 372-382.
[5] K F Tee and E O Ekpiwhre 2018 Reliability analysis and growth curves modelling of fielded road systems World Rev Inter modal Transport Res **7** 2 168-194
[6] Fang Y F, W L Tao and K F Tee 2014 Reliability analysis of repairable k-out-n system from time response under several times stochastic shocks Smart Structural System **14** 4 559-567
[7] Y Zhang, C W Kim, K F Tee and A Garg 2018 Long-term health monitoring for deteriorated bridge structures based on copula theory Smart Structural System **21** 2 171-185
[8] F C Ponzo, R Ditommaso, G Auletta and A Mossucca 2010 A fast method for structural health monitoring of Italian reinforced concrete strategic buildings Bull Earthquake Engineering **8** 1421-1434.
[9] S K Bhattacharyya and A S Cheliyan 2019 Optimization of a sub-sea production system for cost and reliability using its fault tree model Reliability Engineering and System Safety **185** 213-219
[10] J Geis, K Strobel and A Liel 2012 Snow-induced building failures J Perform Constr Facil **26** 377-388
[11] Y F Fang, W Tao, K F Tee 2019 A new computational method for structural reliability with big data Eksploatacja i Niezawodnosc – Maintenance and Reliability **21** 1 159-163
[12] Strobel K and A Liel 2013 Snow load damage to buildings: physical and economic impacts Proc. ICE – Forensic Engineering **166** 116-133

Acknowledgments

The work described in this paper was supported in part by the Foundation from the Excellent Researcher of GUES (G2015003), the Science Foundation, Guizhou, China (Qian Kehe [2018]1055, Qian Jiao Ke He Zi[2017]304, [2016]7063,Qian Jiao He KY[2018]400), Construction of Data Sharing Mechanism and Platform for Precision Poverty Alleviation, Bijie key laboratory of new energy equipment.

Synthesis and characterization of properties of zinc coated Thaumatococcus daniellii extracts for solid state application

Emetere M.E.[1,2] & Ahiara Ikechukwu M.[1],

[1] Department of Physics, Covenant University Canaan land, P.M.B 1023, Ota, Nigeria.
[2] Department of Mechanical Engineering Science, University of Johannesburg, South Africa.

Abstract. Recent technological inventions focus on the eco-friendly device that would not constitute hazard to immediate environment i.e. both in the short and long term. This research focus on screening bio-semiconductor using synthesized zinc coated Thaumatococcus *daniellii* extract. The band gap of sample 1 (extract in methanol solution), sample 2 (extract in ethanol solution) and sample 3 (extract in butanol solution) was calculated as *2* eV, *1.67* eV and *2* eV respectively. Also, the low and high Q-transition energy in the ethanol and butanol solution was able to show that the d-shell (d^{10}) of the zinc atom was modified. This modification is a huge possibility of deriving bio-semiconductor from plant extract.

1. Introduction

Recent research is now focusing on eco-friendly technologies device. The hazards of waste disposal of used electronics device have revealed that the components of electronics are not eco-friendly at the long-term. Hence, there is the need to go on sustainable bio-materials that are eco-friendly both in the short and long term [1-4]. A material that meets this description is the adoption of plant extract. Hence, beyond the use of plant extract for medicinal, pharmaceutical, food additive and dye purposes, it use has extended into electronics such as semiconductor, bio-filters etc.

Cataldi et al.[5] worked on an environmentally friendly, adaptable and applicable electronic component development approach relying on crop components and graphene. A major monomer of sheellac and tomato cuticula with graphene is produced by the combination of a corn-derived protein, aleuritic acid and zein, bio composite conductors with low electrical resistance (almost 10 sq−1) with outstanding mechanical and fatigue strength. Various high-performance electronic applications are observed, such as THz electromagnetic shielding, flexible GHz antenna construction, and flexible solar cell electrode. Comparable to traditional non-degrading equivalents, outstanding performance results are evaluated from each implementation. Azmi et al. [6] did a study on the ability of the extract of aloe vera as an electrolyte for an accumulator; Aloe Vera comprises of acid that can produce electrical current. Results from the experiment identified that aloe vera extract does not have a stable voltage and currency value and demonstrated the potential use of aloe vera extract in reduction the chemicals applied in a conventional electrolyte solution.

In this research, the metallic coated plant extract was synthesized for the purpose of performing initial screening for semiconductor device. The UV-Vis spectrometry was used to characterize the sample. UV-Vis spectrometry is simple and cost-effective characterization method. The UV-Vis light is passed through a sample and the transmission of light by a sample is measured [7-9]. The absorbance is mathematically calculated from the transmittance (T) mathematically by

IOP Conf. Series: Materials Science and Engineering **958** (2020) 012002 doi:10.1088/1757-899X/958/1/012002

$$A = -\log(T) \quad \text{(i)}$$

where transmittance is calculated as

$$T = -\log(\frac{I}{I_o}) \quad \text{(ii)}$$

where I is the intensity of the UV light. The quantity of absorption at any wavelength is due to the molecule's chemical structure [10-12].
Ref [13] used the UV-Vis spectrometry to examine the interactions of molecularly imprinted sensors (MIPs) where only the change in absorbance is measured as a function of wavelength.
Zinc oxide has been proven to have tremendous optical properties that make them a good candidate as optoelectronic devices [14-15].

2. Methodology

Fresh *Thaumatococcus daniellii* leaves (Figure 1) were collected from Ota, Ogun state, Nigeria. The leaves were obtained in less than an hour after rainfall. The inherent chemical components of *Thaumatococcus daniellii* (i.e. tetracontane (28.76%) and L-ascorbic acid (15.07%); hexadecanoic acid (21.62%) and γ-sitosterol (11.06%) [16])

Figure 1. *Thaumatococcus daniellii*

This process is to enable the clean nature of the leave before grinding. The leaves were divided into three portions. Each portion was grinded in electronic blender using methanol, ethanol and butanol as solution for each portion of the leave. The homogenous liquid was filtered using a sieve. The filtrates were placed in enclosed petri dishes and heated in an oven at 90 °C for 1 hour and allowed to cool and settle for 24 hours. After cooling, 0.05 mole of zinc carbonate was added and heated for 20 minutes at 80 °C. The filtrates were stored in airtight bottles. The filtrates were allowed to settle for 24 hours. Then the filtrates characterized using the UV-Vis spectroscopy and water has the diluting medium. The dilution factor is 10.

3. Results and Discussion

The absorbance from the UV-Vis spectrometry is presented in Figure 2. Sample 1 is the zinc coated *Thaumatococcus daniellii* extract that has been processed with methanol. Sample 2 is the zinc coated *Thaumatococcus daniellii* extract that has been processed with ethanol. Sample 3 is the zinc coated *Thaumatococcus daniellii* extract that has been processed with butanol. The zinc wavelength is already known i.e. 362-381 nm [16-17]. In this research, we are more interested in the π-π transitions of the zinc coated plant extract; hence the wavelength that was considered was 400-650 nm. [18-21]. Figure 2a shows sample 1 having one π-π transition at wavelength of 625 nm. This shows that the components of the *Thaumatococcus daniellii* extract in methanol had no influence on the d-shell of the zinc coating. It was observed that the ethanol solution in sample 2 has enabled complex that has enabled the second π-π transitions as presented in Figure 2b. The Zn^{2+} atom that has adjacent singlet and triplet states (that lie on same orbital symmetry) is the likely cause of the second transition. Figure 2c showed the intense Q peak transition energy at 625 nm. The same Q peak transition energy is said

IOP Conf. Series: Materials Science and Engineering **958** (2020) 012002 doi:10.1088/1757-899X/958/1/012002

to be responsible for the transition at 545 nm. Based on literature [22-24], the Q- band of the π-π transitions exists at the point (iii)&(iv) of Figure 2c. The point (i) and (ii) in Figure 2c are as a result of low Q peak transition energy on the components of the *Thaumatococcus daniellii* extract. The transition patterns in samples 1-3 have made easy to identify the Q-transition in the complexes of transition metal ions (e.g. zinc) with macrocyclic ligands (components of *Thaumatococcus daniellii* extract).

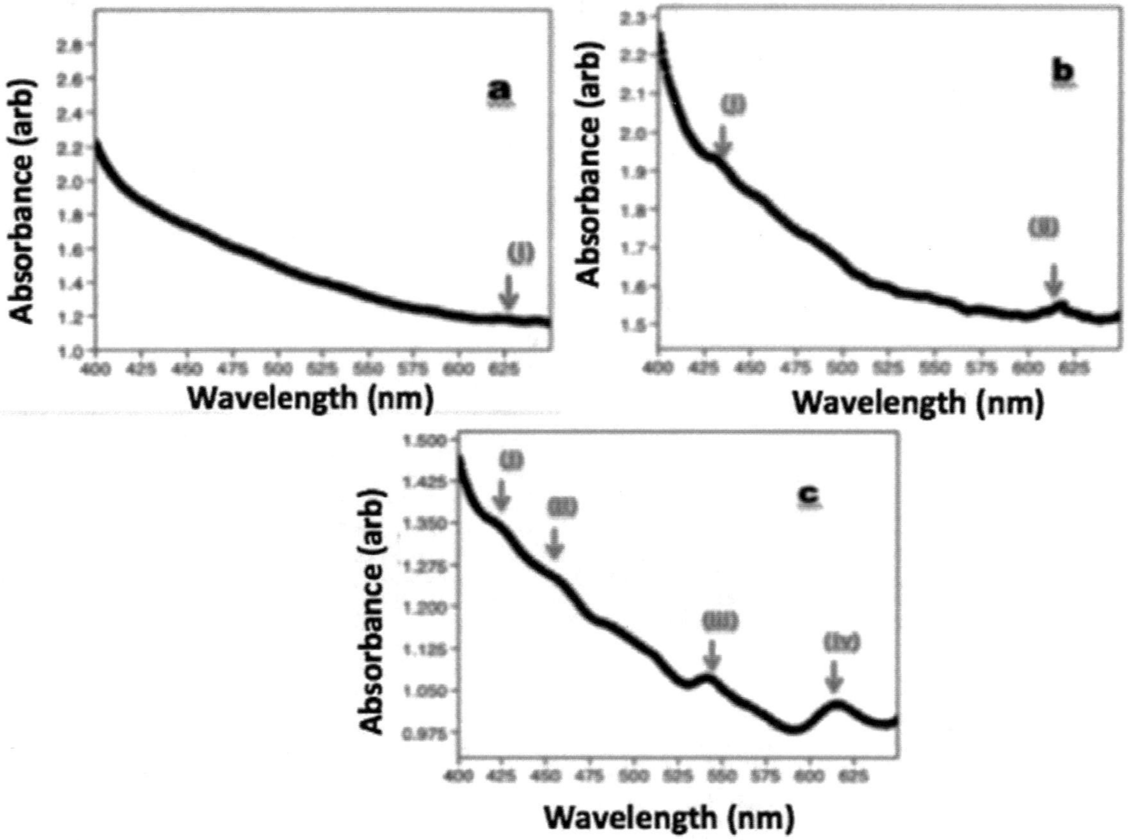

Figure 2. UV-Vis absorption spectrum zinc coated *Thaumatococcus daniellii* extracts (a) extract in methanol solution (b) extract in ethanol solution (c) extract in butanol solution

Using the power series (equation 3) curve fitting, it was easy to identify the coefficients in the equation as presented in Table 1. It can be shown by the coefficients that sample 2 and 3 has something in similar i.e. the low and high Q-transition energy. This is because the ethanol and butanol was able to modify the d-shell (d^{10}) of the zinc atom.

$$y = ax^b + c \qquad \text{(iii)}$$

Table 1. Coefficient of the power series curve fitting

	a	b	c
Sample 1	6.408E5	-2.1638	0.57759
Sample 2	45551	-1.7814	0.99505
Sample 3	22611	-1.7083	0.59194

The statistical models further validated the results in Table 1 as shown in Table 2. The second evidence that sample 2 and 3 has the high and low q-transition energies are shown by the one-way ANOVA. The one- way ANOVA between sample 1 and 2 shows that the following statistical output: variance in the group (sample 1 and 2) is 0.0215395; omega square is 0.1651; Levene´s test for homogeneity of variance, from means p (same) is 2.081E-13; Levene´s test, from medians p (same) is 2.283E-10; Welch F test in the case of unequal variances are F=100.3, df=429, p=2.364E-21.

Table 2. Statistical models for analyzing the absorbance of the synthesized material

	Sample 1	Sample 2	Sample 3
Taxa_S	251	251	251
Individuals	264	268	206
Dominance_D	0,004125	0,00403	0,004039
Simpson_1-D	0,9959	0,996	0,996
Shannon_H	5,508	5,52	5,519
Evenness_e^H/S	0,9831	0,9945	0,9933
Brillouin	3,091	2,675	3,108
Menhinick	13,06	12,23	14,93
Margalef	44,84	44,71	46,92
Equitability_J	0,9969	0,999	0,9988
Fisher_alpha	345,1	261,4	0
Berger-Parker	0,005419	0,004745	0,003536
Chao-1	251	251	252

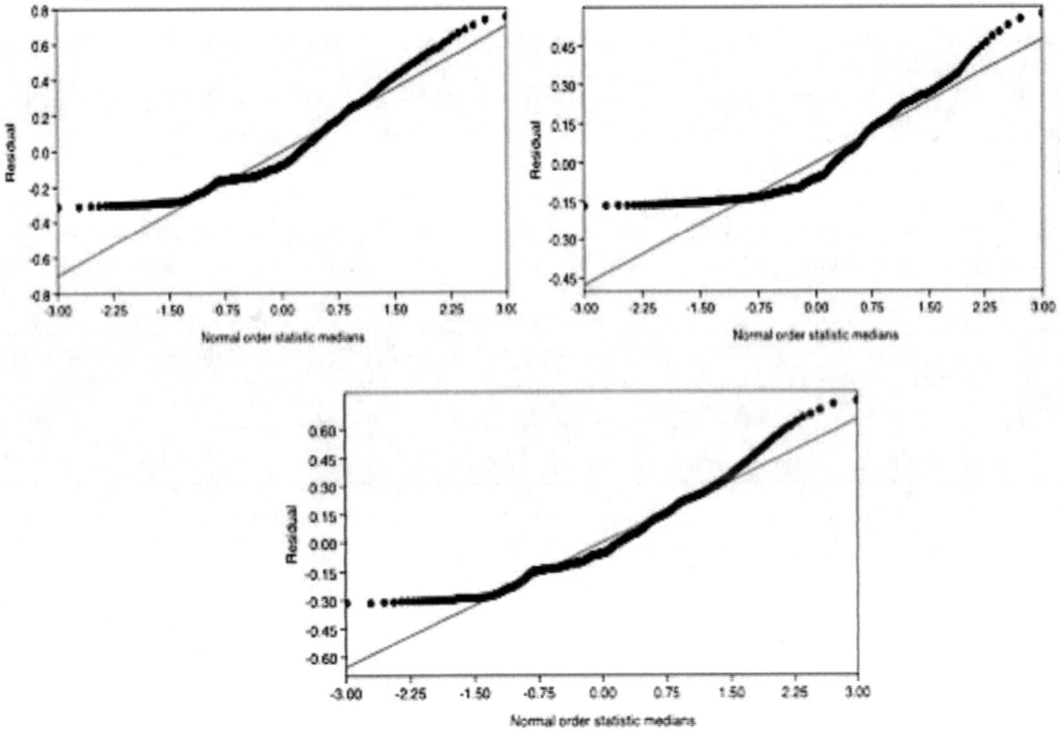

Figure 3. Residual plot of absorption spectrum of zinc coated *Thaumatococcus daniellii* extracts (a) extract in methanol solution (b) extract in ethanol solution (c) extract in butanol solution

The one- way ANOVA between sample 2 and 3 shows that the following statistical output: variance in the group (sample 2 and 3) is 0.152449; omega square is 0.7538; Levene´s test for homogeneity of variance, from means p (same) is 1.877E-06; Levene´s test, from medians p (same) is 0.006585; Welch F test in the case of unequal variances are F=1538, df=459.4, p=1.104E-148. The one- way ANOVA between sample 1 and 3 shows that the following statistical output: variance in the group (sample 1 and 3) is 0.058897; omega square is 0.385; Levene´s test for homogeneity of variance, from means p (same) is 1.123E-28; Levene´s test, from medians p (same) is 1.241E-20; Welch F test in the case of unequal variances are F=315.2, df=358.5, p=4.728E-51. It is clearly shown that sample 2 & 3 had more in common compared to other group. Also, the residual plot presented in Figure 3 affirmed earlier postulation (via the magnitude and pattern of the residuals) that sample 2 and 3 (Figure 3b) is more related. The band gap of the synthesized is given as is given as 2eV, 1.67eV and 2eV for sample 1, 2 & 3 respectively.

4. Conclusion

The zinc coated Thaumatococcus *daniellii* extract has proven to have tremendous properties that makes it a good candidate for the green solution semiconductor device. The band gap of the synthesized is given as is given as 2eV, *1.67*eV and 2eV for sample 1, 2 & 3 respectively. Also, the low and high Q-transition energy in the ethanol and butanol solution was able to show that the d-shell (d^{10}) of the zinc atom was modified. It was discover that the butanol extraction solution (sample 3) was the most viable sample with four π-π transitions that connote that there was formation of oxide layer via the modified adjacent singlet and triplet states that lie on same orbital symmetry.

5. References

[1] Makarov, V. V., Love, A. J., Sinitsyna, O. V., Makarova, S. S., Yaminsky, I. V., Taliansky, M. E., & Kalinina, N. O. (2014). "Green" nanotechnologies: synthesis of metal nanoparticles using plants. Acta naturae. **6**(1):35–44.
[2] N. Ahmad and S. Sharma, (2012), Green synthesis of silver nanoparti- cles using extracts of Ananas comosus, Green and Sustainable Chemistry, **2**(4): 141–147.
[3] Anuoluwa Abimbola Akinsiku, Enock Olugbenga Dare, Kolawole Oluseyi Ajanaku, Olayinka Oyewale Ajani, Joseph Adebisi O Olugbuyiro, Tolutope Oluwasegun Siyanbola, Oluwaseun Ejilude, Moses Eterigho Emetere, (2018). Modeling and Synthesis of Ag and Ag/Ni Allied Bimetallic Nanoparticles by Green Method: Optical and Biological Properties, International journal of biomaterials, 2018:1-14
[4] F.Benakashani,A.R.Allafchian,andS.A.H.Jalali,(2016), Biosynthe- sis of silver nanoparticles using Capparis spinosa L. leaf extract and their antibacterial activity, Karbala International Journal of Modern Science, **2**(4): 251–258.
[5] Cataldi Pietro, José A. Heredia, Guerrero Susana Guzman, Puyol Luca, Ceseracciu Luca, La Notte Andrea ,Reale Jun Ren, Yijie Zhang, Lei Liu, Mario Miscuglio, Patrizia Savi , Simonluca Piazza, Marti Duocastella , Giovanni Perotto, Athanassia Athanassiou, and Ilker S. Bayer (2018), Sustainable Electronics Based on Crop Plant Extracts and Graphene: A “Bioadvantaged” Approach, Advanced Sustainable Systems, 2: 1800069
[6] Azmi F, Sispriatna D, Ikhsan K, Masrur M, Azzahra S.S, Mahidin and Supardan M.D. (2018) Utilization of aloe vera extract as electrolyte for an accumulator. IOP Conf. Series: Materials Science and Engineering, 334 012053 doi:10.1088/1757-899X/334/1/012053
[7] Chen Y, Bagnall D, Yao T (2000). ZnO as a novel photonic material for the UV region. Mater. Sci. Eng.B, **75** 190- 198
[8] Makarov, V. V., Love, A. J., Sinitsyna, O. V., Makarova, S. S., Yaminsky, I. V., Taliansky, M. E., & Kalinina, N. O. (2014). "Green" nanotechnologies: synthesis of metal nanoparticles using plants. Acta naturae. **6**(1):35–44.
[9] Pranlekha T, Kriengkri T, Chatchawal W (2017) Flexible room-temperature resistive humidity sensor based on silver nanoparticles. Mater. Res. Express 4 085038. https://doi.org/10.1088/2053-1591/aa85b6

[10] Paul. J. W., Elias. A.G. Zagatto (2017) Spectrophotometry: Overview. Encyclopedia of Analytical Science (Third Edition), Academic Press, Pages 244-248. https://doi.org/10.1016/B978-0-12-409547-2.14265-9

[11] Libretexts (2017). Ultraviolet and visible spectroscopy, https://chem.libretexts.org/Bookshelves/Organic_Chemistry/Book%3A_Organic_Chemistry_with_a_Biological_Emphasis_(Soderberg)/Chapter_04%3A_Structure_Determination_I/4.4%3A_Ultraviolet_and_visible_spectroscopy (Accessed 27th March, 2019).

[12] Chemistry (2019). UV-Visible Spectroscopy, https://www2.chemistry.msu.edu/faculty/reusch/virttxtjml/spectrpy/uv-vis/uvspec.htm (Accessed 29th March, 2019).

[13] Adnan M., Franz L. D. (2012) Chapter 6 - Molecularly Imprinted Polymers for Sensors: Comparison of Optical and Mass-Sensitive Detection. Molecularly Imprinted Sensors, Elsevier, Pages 125-159. https://doi.org/10.1016/B978-0-444-56331-6.00006-2

[14] Chohan Z. H. , Khan K. M. and Supuran C. T. , (2004), Synthesis of antibacterial and antifungal cobalt (II) , copper (II) , nickel (II) and zinc(II) complexes with bis – (1,′1 – disubstituted ferrocenyl) thiocarbohydrazone and bis-(1,′1 – disubstituted ferrocenyl) carbohydrazone , Appl. Organomet. Chem., **18** 305 – 310,

[15] Soosen Samuel M, Lekshmi Bose and George KC., (2009), Optical properties of ZnO nanoparticles, Academic Review **16** 57-65

[16] A. A. Hamid, M. A. Aliyu, L. Z Abubakar, A. A. Mukadam, A. Shehu, G. Egharevba, M. J. Adisa, S. O. Ajibade, A. O. Zubair, E. O. Fagbohun (2017), Thaumatococcus daniellii leaves: its chemical compositions, antioxidant and antimicrobial activities, Ife Journal of Science, **19**(2):409-416

[17] Richa Khokhra, Bandna Bharti, Heung-No Lee and Rajesh Kumar, (2017), Visible and UV photo-detection in ZnO nanostructured thin films via simple tuning of solution method, Scientific report, **7** 15032

[18] J. Amesz (2000), Photosynthesis, 1st Edition, Elsevier Science, ISBN: 9780080860749, pp. 5

[19] Liao M.-S., Watts J. D., Huang M.-J., Gorun S. M., Kar T., Scheiner S., (2005), Effects of Peripheral Substituents on the Electronic Structure and Properties of Unligated and Ligated Metal Phthalocyanines, Metal = Fe, Co, Zn, J. Chem. Theory Comput., **1** 1201.

[20] Nguyen K. A., Pachter R., (2001), Ground state electronic structures and spectra of zinc complexes of porphyrin, tetraazaporphyrin, tetrabenzoporphyrin, and phthalocyanine: A density functional theory study, J. Chem. Phys., **114** 10757.

[21] Ricciardi G., Rosa A., Baerends E. J., (2001), Ground and Excited States of Zinc Phthalocyanine Studied by Density Functional Methods, J. Phys. Chem. A., **105** 5242.

[22] Djurišić A.B., Kwong C.Y., Lau T.W., Guo W.L., Li E.H., Liu Z.T., Kwok H.S., Lam L.S.M., Chan W.K., (2002), Optical properties of copper phthalocyanine. Opt. Commun., **205** 155–162.

[23] Liu Z.T., Kwok H.S., Djurišić A.B., (2004), the optical functions of metal phthalocyanines. J. Phys. D Appl. Phys., **37** 678–688.

[24] Ottaviano L., di Nardo S., Lozzi L., Passacantando M., Picozzi P., Santucci S., (1997), Thin and ultra-thin films of nickel phthalocyanine grown on highly oriented pyrolitic graphite: An XPS, UHV-AFM and air tapping-mode AFM study. Surf. Sci., **373** 318–332.

Acknowledgement
The authors appreciate Covenant University for partial sponsorship.

Solid State Device from Plant Extracts

Emetere M.E.[1,2] Chijuo C.[1] and Afolalu S.A. [3]

[1]Department of Physics, Covenant University Canaanland, P.M.B 1023, Ota, Nigeria.
[2]Department of Mechanical Engineering Science, University of Johannesburg, South Africa.
[3]Department of Mechanical Engineering, Covenant University Canaan land, Ota, Nigeria

emetere@yahoo.com

Abstract. The infusion of green solution into solid-state technology is timely at a time where land, air, and water pollution (from electronic waste) is high in most parts of the globe. At the moment, several research works have proven that plant extract doped with metallic compound exhibits high electrical, optical and chemical properties. In this research we examine two different plant extracts with different dopant. The special opto-transport property of the synthesized compounds was examined using an unconventional characterization technique. It was observed that both compounds can be identified as solid-state device with remarkable optical properties.

1. Introduction

Solid-state devices are systems or electronic components solely dependent on a semiconductor. They allow electricity to flow through a solid semiconductor crystal like silicon, germanium, and gallium arsenide. In recent times, the definition of solid-state becomes more interesting with the emergence of different material in solid and liquid state. Other than the normal explanations given for solid-state devices, it is also often described as semiconductor electronics or electronic equipment that uses devices like transistor, integrated circuits and others. Other example of solid-state device includes LED lamps, solar cells, CCD image sensor that many people use in cameras and semiconductor laser [1-3]. Another application of solid-state devices is in computers i.e., in form of component or storage device.
Semiconductors are known to have electrical conductivity between that of an insulator and that of most metals, either due to the addition of an impurity or because of temperature effects. Examples of the most famous semiconductors used nowadays are the Transistor and the Junction diode [4-5]. They are electronic device that works due to the movement of electrons in a solid piece of semiconductor material. The emergence of liquid semiconductors has proven that the physics in solids can be replicated in liquids. Liquid solid-state device includes hydrogenated amorphous silicon and mixtures of arsenic, selenium and tellurium in a variety of proportions. They are generally used in thin film structures, which do not require material of higher electronic quality, Since the creation of the transistor; solid-state devices have developed at a wonderful rate [6]. Maximum efforts have been put towards the methods they are made and there is no limit to the devices in future.

The idea of green solution into electronics was borne via the massive pollution from electronic wastes. Several deaths and diseases are reported all over the globe from electronic waste [7]. The rate of death is higher in developing countries where there are no regulation guiding waste disposals. Despite there are laws to guide the production of solid-states device or electronics gadgets, the long term danger of controlling its waste still subsists. The idea of using plant extracts became prominent because it believed that the plant wastes are mostly biodegradable. Also, some plants contain vital chemical components that make it a good candidate as semiconductor. For example, garlic is known for being rich in both germanium and selenium. Now the content of selenium alone is very dependent on its availability in the soil and this is based solely on the location. The presence of these two elements is very high in garlic compared to some fruit and vegetable like potato, carrot, apple etc. According to research conducted by notable scientists and researchers worldwide, it has been found that both of them possess contents of about 2.78 and 0.276 $\mu g\ g^{-1}$ respectively in garlic [8]. Some of these plants are mined to get certain chemical elements or compounds. During plant mining, some chemicals are used for the extraction process for different concentration and quantity of the desired elements [9-11]. The final product certainly would not be regarded as biodegradable.

The use of unprocessed plant extract in itself is frustrating because plant contains hydrogen and carbon in disturbing quantities that makes it an insulating liquid. The idea of doping the extract with certain quantities of impurities or metallic compounds is to allow for the chemical engagements of some of the redundant chemicals in plants [7]. The bonding or antibonding is more prominent and can be varied based on the concentration of the dopants. However, the quantity of the dopant is very important, as the basic aim of this experiment is to avoid long-term electronic wastes. In this study, the two different plants extract with different dopants were analyzed determine its opto-lectrical properties [12-13].

2. Methodology

Ixora plants are small shrubs that grow in subtropical regions. These plants feature large clusters of red, yellow, white, or orange flowers that emerge like puffballs from the evergreen leaves. The phytochemical screening of ixora plant reveals the following compounds: phenols, alkaloids, flavonoids, terepnoids, coumarins, tannin, saponin, anthocynin, anthraquinone and amino acids [14]. Flavonoid is found to be the main chemical in the ixora plant.

The flow chart of the methodology is presented in Figure 1. The flower was rinsed in water and placed in a filter to drip its water over night. The flower was grinded using methanol alone. The filtrate was filtered to obtain a red extract. The extract was doped with 0.01M of copper nitrates compound and thoroughly mixed. The doped plant extract was allowed to react under room temperature and pressure for a day.

The second method is to characterize the liquid semiconductor using unconventional characterization technique to test the opto-transport properties of the doped plant extract (Figure 2). The doped plant extract was sprayed on a photovoltaic (PV) to monitor its opto-transport.

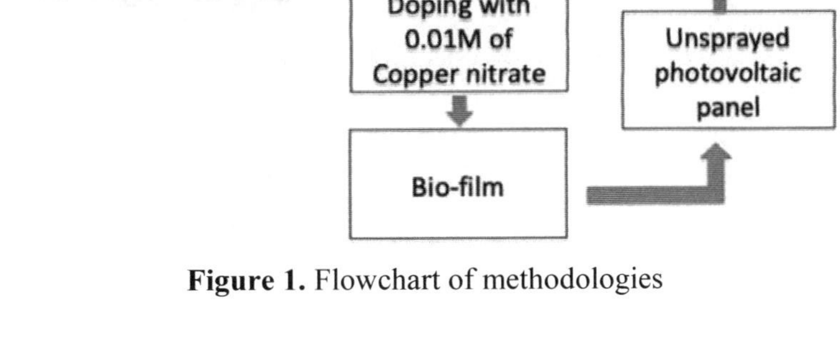

Figure 1. Flowchart of methodologies

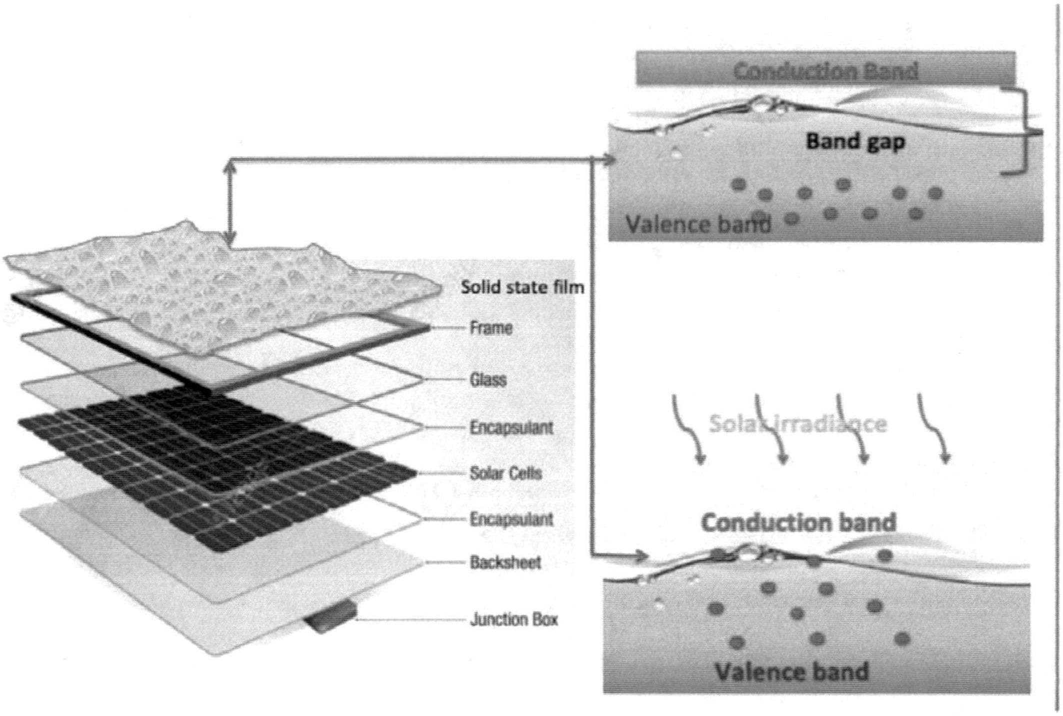

Figure 2. Physics of unconventional characterization technique

A clean photovoltaic panel that is placed at the same time with the sprayed PV to remotely monitor the physics of the doped plant extract. The opto-transport is monitored by the differing voltage output of the both PV panels. The solid-state film (doped plant extract) is expected to react first before the solar irradiance passes to the PV panel. The opto-transport of the solid state film can be determined by the equations below.

$$A = -\log\frac{I}{Io} = -\log\left(\frac{V_{sample} - V_{zero}}{V_{solvent} - V_{zero}}\right) \tag{1}$$

Where A is the absorption, V_{sample} is the voltage on the clean PV module; $V_{solvent}$ is the voltage of the sprayed PV module. V_{zero} is the differential voltages between the sprayed and unsprayed PV module.

3. Results and Discussion

Ixora coccinea flower is of ursolic acid chemotype [15]. The formular of the ursolic acid is $C_{30}H_{48}O_3$. After the compound was doped, the chemical reaction is

$33\ H_2O + 64\ CuNO_3 + 7\ C_{30}H_{48}O_3 = 64\ CuNC_3H_6O_3 + 18\ HCO_3$.

The XRD was examined at wavelength 1.54056 nm. The XRD results show that the compound is polycrystalline with copper. The copper atom is the highest peak with plane and d-spacing as (002) and 8.493Ã. The nitrogen element is on plane (012) and d-spacing of 7.17Ã; carbon element is on planes (100), (022), (110) and with d-spacing of 5.41 Ã, 5.26 Ã, 5.02 Ã respectively. The hydrogen element is on planes (10-2), (11-2), (121) and with d-spacing of 4.65 Ã, 4.40 Ã, 4.06 Ã respectively. The remaining peaks are the oxides and the unidentified peaks. The second type of characterization is the unconventional techniques presented in Figure 4.

IOP Conf. Series: Materials Science and Engineering **958** (2020) 012003 doi:10.1088/1757-899X/958/1/012003

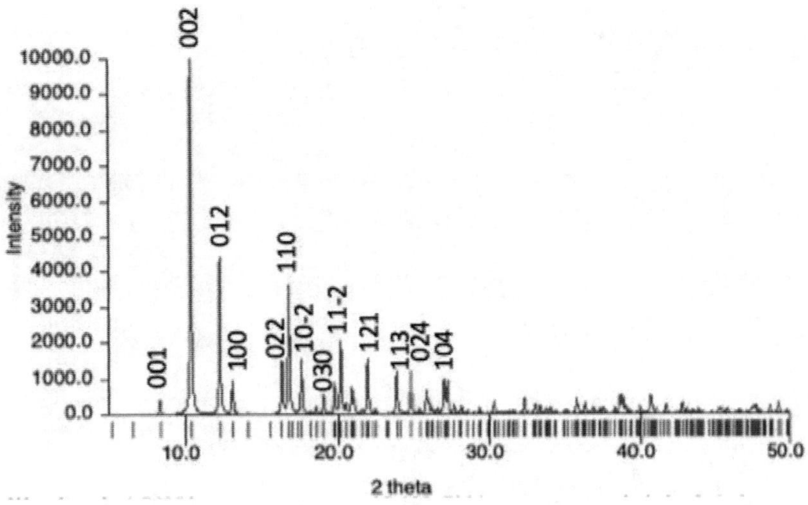

Figure 3. XRD of doped plant extract

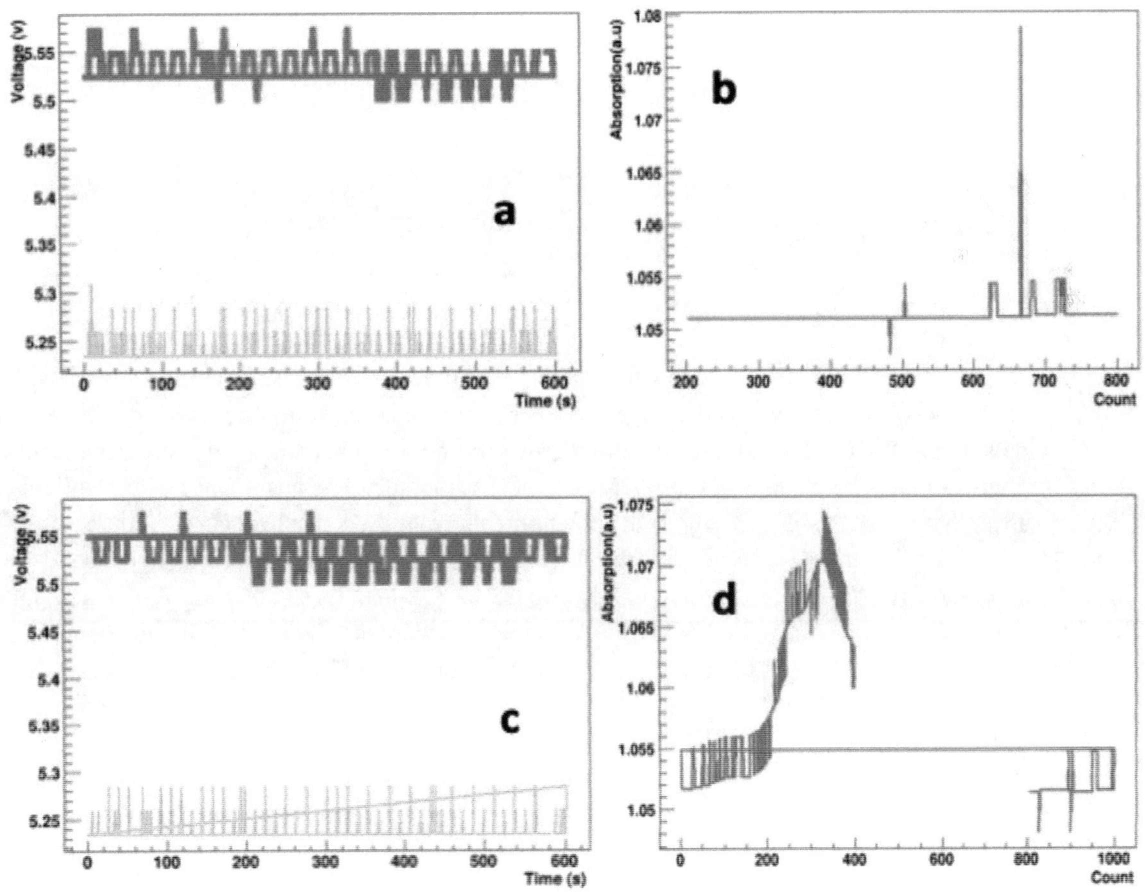

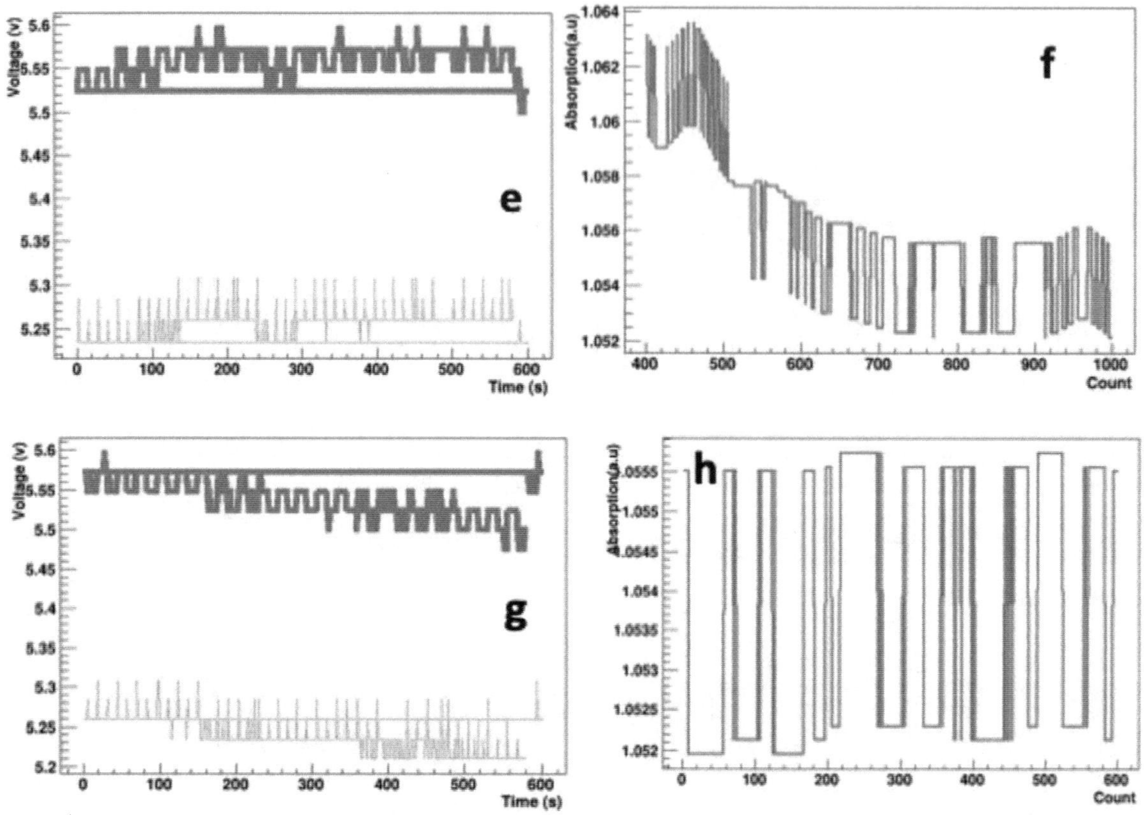

Figure 4. Unconventional characterization technique of the doped plant extract

The red lines in Figures 4a,c, e & g represents the voltage from unsprayed PV module while green lines are in Figures 4b, d, f and h represents the voltage from the sprayed PV module. The features of the peaks in the unsprayed PV modules are in lines and curves, while the voltages in the sprayed PV module are in peaks. This is because the doped plant extract is a liquid semiconductor that converts the original solar signals on the PV module. The different patterns shown in Figures 4a,c, e & g are dependent on solar irradiances that are incident on the surface of the PV module. The absorbance of the solid-state device shows that it reacts significantly to different signals from the solar irradiance (Figures 4b, d, f and h). When there are more curves above the 'line of average', it gives normal peaks in the solid-state device and the absorbance results. In other words, Figure 4 a & b is in its stable state. When there more curves below the 'line of average', the system is unstable and there are likely cases of anti-bonding i.e. judging by the crossed line over the peaks (green line in Figure 4c). In this case, the absorbance shows that the solid-state slightly perturbed. Figures 4e & f shows that the sectional transient states on the doped plant extract and the solar cells in the PV module. At harsh solar irradiances, the absorbance of the doped plants is tremendously controlled, i.e. making it a very good solid-state device.

4. Conclusion

In conclusion, it is wise to note that these solid-state device found in the doped plant extracts are advantageous for safety of humanity by reducing death from e-wastes pollution and for economical purposes. This invention will become a great source of value to the electronic world considering the fact that they have good opto-transport properties to act as bio-filter to shield PV modules from harsh weather. Also, the new solid-state device shows that it can be upgraded into switching device such as metal–oxide–semiconductor field-effect transistor (MOSFET). There are other verse uses of the new solid state.

The unconventional characterization technique has shown to be very effective in determining the opto-transport properties of the new solid-state device. It is recommended that the molecular dynamics of the new solid-state device be investigated to see the best way to optimize the new material.

References

[1]. Ravindranath Tagore Yadlapalli, Rajanand Patnaik Narasipuram and Anuradha Kotapati (2019), An overview of energy efficient solid state LED driver topologies, Int. J. of Energy Research, **44** 612-630

[2]. Slocum, A. A Technology Assessment of Light Emitting Diode (LED) Solid State Lighting for General Illumination; National Center for Environmental Economics (NCEE): Washington, DC, USA, 2005.

[3]. Moses E. Emetere, Marvel L. Akinyemi, and Etimbuk B. Edeghe, (2016). A Simple Technique for Sustaining Solar Energy Production in Active Convective Coastal Regions, International Journal of Photoenergy 2016, 3567502, 1-11, http://dx.doi.org/10.1155/2016/3567502

[4]. G. Wang and G. C. M. Meijer, (2000). The temperature characteristics of bipolar transistors fabricated in CMOS technology, Sens. Actuators A, **87** 81–89,

[5]. Ken Uchida, Jugli Kaga, Ryuji Ohba and Akira Toriumi, "Programmable Single-Electron Transistor Logic for Future Low-Power Intelligent LSI: Proposal and Room-Temperature Operation", IEEE Transactions on Electron Devices, **50** (7), July 2003.

[6]. E. Fortunato, P. Barquinha, R. Martins (2012), Oxide semiconductor thin-film transistors: a review of recent advances, Adv. Mater., **24** 2945-2986

[7]. Emetere Moses E., Adeyemo Nehemiah, (2019). Spectral Filtering of Photovoltaic Cells using Novel Bio-Filter: Silver Coated Hibiscus Extract using Butanol Solution, International Journal of Engineering and Advanced Technology, **9** 724-727

[8]. A.F. Koca, I. Koca, B. Tekguler, (2016). Two antoxidant elements of Allium vegetables: germanium and selenium, Acta Hortic., **1143**. 297-302

[9]. Kalisz, A.; Sekara, A.; Smoleń, S.; Grabowska, A.; Gil, J.; Komorowska, M.; Kunicki, E. (2019), Survey of 17 elements, including rare earth elements, in chilled and non-chilled cauliflower cultivars. Sci. Rep., **9** 5416.

[10]. Sium M, Kareru P, Keriko J, Girmay B, Medhanie G & Debretsion S (2016). Profile of trace elements in selected medicinal plants used for the treatment of diabetes in Eritrea. The Scientific World Journal, 2016.

[11]. Z. Ding, Y. Li, Q. Sun, H. Zhang (2018), Trace elements in soils and selected agricultural plants in the Tongling mining area of China, Int. J. Environ. Res. Public Health, **15** 202

[12]. M.A. Islam, A.M.M. Tanveer Karim, M. Julkarnain, A.K.M. Badrul, K.A. Khan (2017). Opto-transport properties of e-beam evaporated annealed CuInSe2 thin films, Surf Interf, **8** 170-175

[13]. M. Younus Ali, M.K.R. Khan, A.M.M. Tanveer Karim, M. MoziburRahman, M.Kamruzzaman (2020). Effect of Ni doping on structure, morphology and opto-transport properties of spray pyrolised ZnO nano-fiber, 6: e03588

[14]. Sneha G. Nair, Vikram R. Jadhav & Satyam S. Bakare (2018), Ixora coccinea: Study of Phytochemical Parameters and Antioxidant Activity, International Journal of Innovative Research in Science,Engineering and Technology, **8** (7): 9051-9064

[15]. A.Elumalai, Chinna Eswaraiah, Yetcharla Venkatesh, Burle Shiva kumar and Chava Narendar (2012)., Phytochemical and pharmacological profile Of Ixora coccinea Linn. International Journal of Pharmacy & Life Sciences, 3: 1563-156

Acknowledgment

The authors appreciate the partial sponsorship of Covenant University.

IOP Conf. Series: Materials Science and Engineering **958** (2020) 012004 doi:10.1088/1757-899X/958/1/012004

Effects of Thermal Diffusivity Analysis after Irradiation

Ahli K. D. Willie[1, a *], Hongtao Zhao[2,b] and Teplinskaya Svetlana3,c

[1]Fundamental Science on Nuclear Safety and Simulation Technology Laboratory, Harbin Engineering University, Harbin, 150001, China
[2]Institute of Technical Physics, Heilongjiang Academy of Sciences 150086, China
[3]College of Material Science and Chemical Engineering, Harbin Engineering University, Harbin, 150001, China

willie@hrbeu.edu.cn

Abstract. The diffusion calculation gives a vivid understanding as to what happens in the SiC-cladded material. Molecular Dynamics (MD) and Molecular Statics are being employed to study the diffusion coefficient phenomena. The MD simulations in this study are been built on the ZBL potential. In this work we initially applied the MD simulation for minimization within the temperature range of 1000–3000 K. Then the MOX fuel is then used to perform assessment of radiation damage by ions at burnup temperatures as well. Various chemical states are developed depending on the condition of the fuel. Within the fuel lattice the O atoms break bonds with the U-Pu atoms at higher temperature. The very short diffusion lengths mechanisms results were obtained measured for uranium atom over the course of this 300ps simulation.

1. Introduction

All Nuclear fuel raw materials come from the earth like the fossil fuels. These major materials generate its radioactive by-products from the extraction, enrichment, fabrication, and consumption stages of the process [1]. When a spent fuel is taken out of the reactor it produces a lot of heat as well as radiation. Prior to previous studies metallurgy been used to enhance mechanical properties which have shown that adding a duration quantities with several oxides to ceramic bodies will have large effects on the mechanical performance and microstructure of the ceramic [2-4]. During the inclusion of minor actinides like CmO2, NpO2 and AmO2 with UO2 or ThO2 is desirable so that these species can undergo transmutation in a reactor [5]. Computational models can be used to analyze the nuclear materials so that we can have a better understanding of their behaviors, which can help to increase their efficiency and stability.

Silicon carbide (SiC) has an excellent thermo-mechanical property that makes it a subject of interest for high-temperature mechanical applications. There are numerous phases with high thermal conductivity, including the silicon carbide nanoparticles SiC [6][7] and nanotubes [8], which when added to UO2 pellets makes thermal conductivity improved. However, there are two main conflicting effects as to when introducing the SiC into UO2. Initially, with a high thermal conductivity substance, SiC will enhance the thermal conductivity of UO2. Nevertheless, as a strong-covalent-bond compound, SiC increases the sintering difficulty, reduces the density and consequently weakens the thermal conductivity. Therefore, there shall be a critical sintering condition at which the enhancing effect counteracts the weakening effect.

IOP Conf. Series: Materials Science and Engineering **958** (2020) 012004 doi:10.1088/1757-899X/958/1/012004

In this present work, additional calculations and MD simulations were employed to completely evaluate the impact diffusion of (SiC, U-Pu) O2 microstructural-based system. This is very significant towards understanding results of non-uniform lattice change of the mixed oxides from no to high burnup. This affirms the diffusion mechanisms exclusively have a dependent factor on point defects. Emphasis is place on comparing excessive oxygen diffusion coefficient through the comparison in with the mixed oxide fuel. This was ideally made so that predictions can begin to be made for actinide oxides.

2. Interatomic Potential Model

Molecular dynamics simulation provides a unique insight with the atomic-scale processes applicable to solve problems in material science. We performed our simulation by using the Large-scale Atomic/Molecular Massively Parallel Simulator (LAMMPS) code from Sandia National Laboratories [9]. Molecular dynamics simulation uses a force computed in empirical potentials, usually known as classical molecular dynamics.

The interatomic potential energy of a system $V(r)$ of n atoms with coordinates r_i, r_j, …, r_n can be written in terms of one, two, as well as many-body terms as:

$$V(r) = \sum_{i}^{n} V_1(r_i) + \sum_{i}^{n}\sum_{j>i}^{n} V_2(r_i, r_j) + \sum_{i}^{n}\sum_{j>i}^{n}\sum_{k>j>i}^{n} V_2(r_i, r_j, r_k) \tag{1}$$

Furthermore repulsive potential described in the short interatomic distances is accustomed to collision cascade simulations as a result of high kinetic energy within atoms. This geometry optimization was carried out in LAMMPS using the empirical potential of nuclear fuel model Ziegler–Biersack–Littmark (ZBL) [10].

$$V(r) \approx \sum_{i}^{n} V_1(r_i) + \sum_{i}^{n}\sum_{j>i}^{n} V_2^{eff}(r_{ij}) \quad with\, r_{ij} = |r_i - r_j| \tag{2}$$

This effective potential $V_2^{eff}(r_{ij})$ perhaps is disintegrated into the long-range Coulombic interaction of ionic particles and the short-range portion.

$$V_2^{eff}(r_{ij}) = V_{LR}(r_{ij}) + V_{SR}(r_{ij}) \tag{3}$$

Within the charges q_i and q_j distant from each other by r_{ij}, the long-range portion $V_{LR}(r_{ij})$ is:

$$V_{LR}(r_{ij}) = \frac{q_i q_j}{r_{ij}} \tag{4}$$

The interactions with each other and with U-Pu and O typically involved, produces only short-range repulsion and dispersion interactions [11]. The short-range portion can be written as:

$$V_{SR}(r_{ij}) = A_{ij}e^{-B_{ij}r_{ij}} - \frac{C_{ij}}{r_{ij}^6} + D_{ij}\left[e^{-2\beta_{ij}(r_{ij}-r_{ij}^0)} - 2e^{-\beta_{ij}(r_{ij}-r_{ij}^0)}\right] \tag{5}$$

IOP Conf. Series: Materials Science and Engineering **958** (2020) 012004 doi:10.1088/1757-899X/958/1/012004

$$V_2^{eff}(r_{ij}) = \frac{q_i q_j}{r_{ij}} + A_{ij}e^{-B_{ij}r_{ij}} - \frac{C_{ij}}{r_{ij}^6} + D_{ij}\left[e^{-2\beta_{ij}(r_{ij}-r_{ij}^0)} - 2e^{-\beta_{ij}(r_{ij}-r_{ij}^0)}\right] \quad (6)$$

$$V_{ZBL}(x) = 0.1818e^{-3.2x} + 0.5099e^{-0.9423x} + 0.2802e^{-0.4029x} + 0.02817e^{-0.2016x} \quad (7)$$

With $x = \frac{r_{ij}(Z_i^{0.23} + Z_j^{0.23})}{0.8854a_0}$

Z_i and Z_j representing atomic numbers and a_0 the Bohr radius.

All time step of a molecular dynamics (MD) run has an optimisation of the shells to their zero-force positions for each atomic configuration. Short-range interatomic potential has effect on the shell particles which will then decrease some computational activity within the non-bonded interactions. There is no electrostatic force of interaction that exists between the core as well as its own shell. The shell models give a better representation of the equilibrium properties including the elastic constants in static calculations [12].

The Density functional theory (DFT) data was obtained for the selected MOX property of interest using Molecular statics calculations, such as melting temperature, oxygen vacancy migration energy and diffusion coefficient phenomena. The interaction potentials valid for all values of r_{ij} have been developed for (SiC, U-Pu) O_2 by fitting to a database of energies of different structures calculated using DFT [13]. This was done due to reduce any uncertainty produced by the selected functional form and range of activity within the spline.

3. Results and Discussion

The simulation results revealed several derived important aspects using the diffusion coefficients for large and small (SiC, U-Pu) O_2 clusters. The results seems to be trailing off towards lower temperatures, which is clearly coherent, since an error or noise could result in an over estimation of the diffusion coefficient when the number of displacement is small. As can be observed in Fig. 1 the derived diffusion coefficient for the small 324-ion (SiC, U-Pu) O_2 cluster. The fitted graphs correspond with 0.53 eV and 0.31 eV of activation energy for diffusion.

In Fig. 2, a representative plot of the diffusion coefficients of oxygen derived from 3129 ion with (SiC, U-Pu) O_2 cluster results. Here the surface migration results are clearly invalid. The bulk oxygen results show almost similar behavior as the 324-ion cluster, where the diffusion coefficient seems higher than expected at lower temperatures. Although the diffusion length of uranium is almost certainly too short to perform any reliable analysis derived diffusion coefficients which is presented in Fig. 3. The widening of the distribution graphs due to thermal motion of the ions is what is calculated.

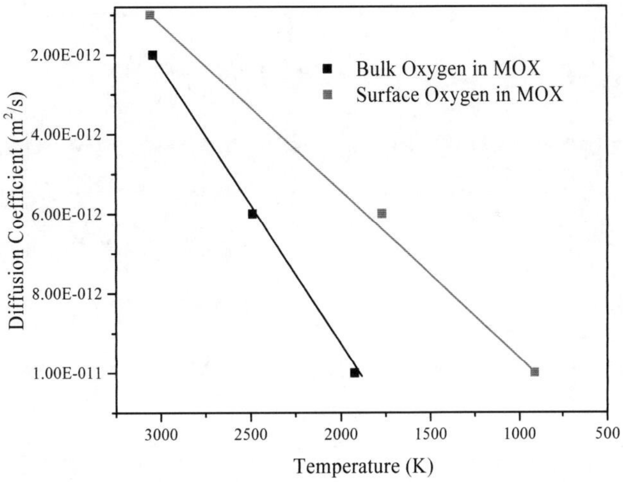

Figure 1. Analyzed oxygen diffusion coefficients derived from the small 324 ions (SiC, U-Pu) O_2 cluster.

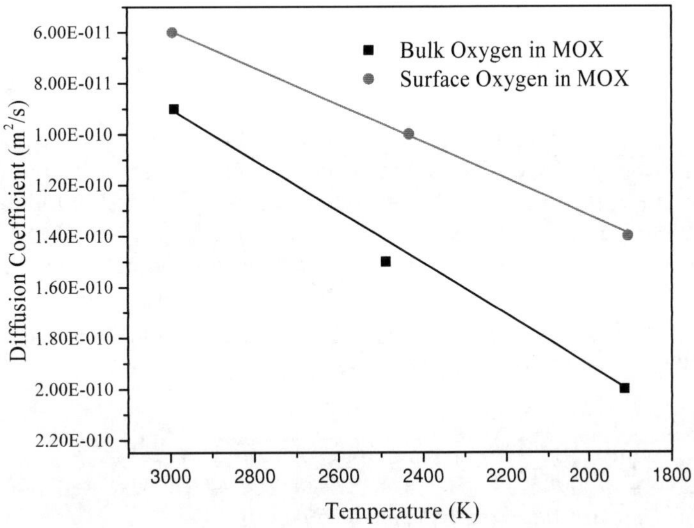

Figure 2. Analyzed oxygen diffusion coefficients derived from the large 3129 ions (SiC, U-Pu) O_2 cluster.

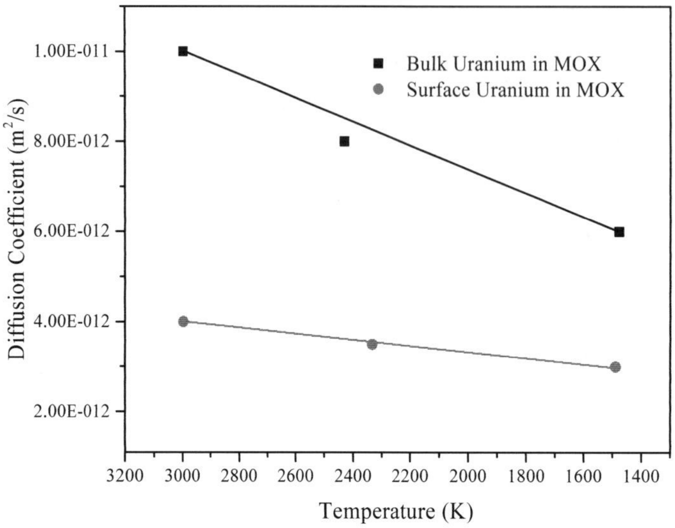

Figure 3. Analyzed uranium diffusion coefficients derived from the large 3129 ion (SiC, U-Pu) O_2 cluster.

4. Conclusion

In this work, a pair in the simulated structure with the interatomic ZBL potential function was used. With the ZBL potential we plotted a comparison of the LAMMPS calculated diffusion coefficients of the MOX with temperature dependence. These graphs could be seen that lattice predictions for the pre-exponential factor were all in the range of 10^{-12} to 10^{-10} m^2/s. A crystal of 3129 ions of (SiC, U-Pu) O_2 cluster was simulated at various temperatures. The crystallite which was used during this simulation was cut along the stable (111) plane. This was done to obtain pure stoichiometric MOX. The influence of SiC is more significant in the MOX having results which further verify the developed formula. This model presented can be incorporated into fuel recycling model to improve calculations of the fuel accident conditions and thermal conductivity as well.

References

[1] J. Bruno, and R. C. Ewing, (2006) Elements 2 343.
[2] I. Amato, R. L. Colombo and A. P. Balzari, "Grain growth in pure and titania-doped uranium dioxide," Journal of Nuclear Materials, **18** (1966) 252-260.
[3] J. Arborelius, "Advanced doped UO2 pellets in LWR applications," Journal of Nuclear Science and Technology, **43** (9) (2006) 967-976.
[4] S. Kashibe and K. Une, "Effects of additives (Cr2O3, Al2O3, SiO2, MgO) on diffusional release of 133-Xe from UO2 fuels," Journal of Nuclear Materials, **254** 234-242, 1998.
[5] Y. I Chang, P. J Finck, and C. Grandy, "Advanced Burner Test Reactor Preconceptual Design Report" (2006) ANL Technical report, Argonne.
[6] S. Yeo, E. Mckenna, R. Baney, G. Subhash, J. Tulenko, Enhanced thermal conductivity of uranium dioxide–silicon carbide composite fuel pellets prepared by Spark Plasma Sintering (SPS), J. Nucl. Mater. **433** (2013) 66–73.
[7] R. Hunt, G. Silva, T. Lindemer, K. Anderson, J. Collins, Preparation of uranium fuel kernels with silicon carbide nanoparticles using the internal gelation process, J. Nucl. Mater. **427** (2012) 245–248.
[8] A. Cartas, H. Wang, G. Subhash, R. Baney, J. Tulenko, Influence of carbon nanotube dispersion in UO2–Carbon nanotube ceramic matrix composites utilizing spark plasma sintering, Nucl. Technol. **189** (2015) 258–267.

[9] S. Plimpton, “Fast Parallel Algorithms for Short-Range Molecular Dynamics,” J. Comput. Phys., **117** (1) (1995) 1–19.
[10] J. F. Ziegler, J. P. Biersack, U. Littmark, and H. H. Anderson, The stopping and ranges of ions in matter. Vol 1 ; The stopping and range of ions in solids, by J.F. Ziegler, J.P. Biersack and U. Littmark Vol 6 ; Handbook of range distributions for energetic ions in all elements, by U. Littmark and J.F. Ziegler. Vo. New York: Pergamon, 1985.
[11] H. Y. Geng, Y. Chen, Y. Kaneta, M. Kinoshita, “Molecular dynamics study on planar clustering of xenon in UO2”, Journal of Alloys and Compounds, (2007) 457, 465.
[12] K. Govers, S Lemehov, M. Hou, & M. Verwerft, “Comparison of interatomic potentials for UO2. Part I: Static calculations”. Journal of Nuclear Materials, **366**(1-2), (2007) 161-177.
[13] P. Tiwary, A. Walle, B. Jeon, and N. Grønbech-Jensen, “Interatomic potentials for mixed oxide and advanced nuclear fuels”, Physical Review B, **83**, 094104. (2011)

Acknowledgments

This paper is funded by the International Exchange Program of Harbin Engineering University for Innovation-oriented Talents Cultivation.

IOP Conf. Series: Materials Science and Engineering **958** (2020) 012005 doi:10.1088/1757-899X/958/1/012005

Learnability Factors for Investigating the Effectiveness of Augmented Reality Smart Glasses in Smart Campus

Sayera Hafsa, Mazlina Abdul Majid*

Universiti Malaysia Pahang, Malaysia.
Faculty of Computing, College of Computing and Applied Sciences, Universiti Malaysia Pahang, Malaysia.

mazlina@ump.edu.my

Abstract. Augmented Reality (AR) has become an emerging platform in the field of learning either in education or industry. Wearable AR smart glasses are the example of new wearable technology devices in AR. AR smart glasses has been introduced as another alternative to enhance the user's experience in the real world rather than replacing the principle of learning. In education, AR smart glasses is seen as a possible technology to be embedded in the teaching and learning among lecturers and students for achieving smart campus environment. The effectiveness of AR smart glasses in education is a prospect for further study. Thus, this paper aims to identify the factors for investigating the learnability of students while learning using AR smart glasses. Various existing learning factors were reviewed through literature exploration and potential AR smart glasses learning factors are suggested. The results are significance for the development of the AR smart glasses learnability model that will further investigate in the future work.

1. Introduction

Augmented Reality (AR*)* is in the information technology industry for a long time and it has emerged from a science-fiction phenomenon to a science-based reality. AR is an insight of the real physical world where we can find that the elements present in and around us are improved and augmented by computer-generated feedback that includes audio, video, *GPS* graphics which overlays the real inputs and many more things have changed in the new digital era. Additionally, AR has become a reality. AR is available in various forms such as on the smartphones, smart glasses, head-mounted displays which will give the augmented display straight in front of the eye, Face Tracking etc. and many other AR devices. Microsoft HoloLens and Google glass may become a stream of breaking new technology of wearable devices referred to as augmented reality smart glasses which may well influence media usage shortly [1].

Informatics has digitized human beings with the contribution made by the AR in the education sectors to enhance learnability, AR is the subarea of computer science it provides us with hardware and software dedicated to present the virtual world, or cyber world onto the real world which facilitates the educational sectors and learning to a greater extent. Augmented reality has enhanced the education industry in fact it has revolutionized this industry with many unique features, for example the medical students can have a virtual body available right in front of them for practicing the surgical procedures instead of the actual body which can be difficult to have at many situations. Automatically, student experience can be increased by using AR and such experience is one of the main factor to implement smart campus [2], Thus this paper is intended to discuss about factors for investigating the learnability

IOP Conf. Series: Materials Science and Engineering **958** (2020) 012005 doi:10.1088/1757-899X/958/1/012005

of students while learning using AR smart glasses. The work is about reviewing the existing learnability factors in the Literature Review, and next identifying the initial learnability factors for AR smart glasses.

2. Literature Review

The *AR* smart glasses usage has been disclosed by many top information technology companies such as Microsoft, Facebook and Google for the future movement in mass media technology it is where it can be decide to use AR wearable glasses. According to the market predictions, it is the most expected phenomenon that these augmented reality smart glasses are predicted to be successive ‘big thing’ with the development of the upcoming media trends as forecasted in Technavio (2015).

Nowadays AR has emerged as a vital technology that is needed extensively in various areas and most importantly in educational smart campus environments. In recent times the AR usage has increased so extensively in the educational and training areas [3]., also due to the fact that it does not need the support of high-priced hardware and sophisticated instruments, Furthermore the employment of good and light weighted smart glasses have currently reached the masses, Moreover the advancement in augmented reality have paved the path for the adoption of this type of technology on overall basis within the educational organizations in order to enhance the learnability in the smart campus.

Factual study reveals the importance of various drivers or factors like practical advantages, simple use, individual distinction, adaptable factor, whole attitudes, and accepted behavior, though these glasses are put on equally for style fixtures and to capture numerous individual data as well, personality demonstration advantages and impending confidentiality clauses appear less probably to impact these glasses acceptance [1]. The research is from the perspective of the teaching and learning it additionally depicts the motivations to use these kinds of glasses in line with the would-be operators and its study also helps to inspire increased *AR* wearable smart glasses usage.

Philipp A. Rauschnabel [4, 15] presented a view point of would-be operators and focussed on motivators and obstacles of Smart glasses, Moreover, it also focusses on aspects such as indulgent (related to desire and pleasure) and social aspects, whereas Murat Akcayir, Gokce Akcayir [5]. The author focused on blessings and challenges related to *AR* from completely a different perspective. However, the author Apurva and others focused from the teaching aspects and based their studies for analysis by finding within the dynamics impacting of the acceptance of AR devices [6, 16]. Furthermore Bacca and other authors [7] review studies with a spotlight based on work factors like the utilizations, points of interest, restrictions, adequacy, difficulties, and choices of AR in instructional and teaching settings, In addition to all these the author Tien-Chi focused his research on the emotional effect of assimilating augmented reality in teaching [8].

On the social control front, the study by Caudell, T. P, & Mizell, D.W provides valued intuitions for the usage of the AR devices [9].However, smart glasses and apps ought to be able to fulfil consumer’s wants the factors that are taken into consideration throughout his study are purposeful advantages, easy use, individual attention, complete attitudes, and social norms, The analysis methodology used relies on a survey done on the sample from Deutschland [9]., Though it has shown ample positive factors associated with it. However, it does have some limitations like sample restricted to the country Deutschland omitted hedonistic and social aspects and additional analysis is done on user half regarding the appliance aspect.

All the above researchers have dealt with AR wearable devices or smart glasses in areas such as adoption of AR, emotional effect of AR and tests or benefits of AR. However, not enough analysis and research is carried out within the field of effectiveness of AR technology by using the smart glasses devices. Such gap is vital in understanding how the AR smart glasses can enhance the learnability of people who use them. Yet, there is still one question remain “What are the factors to consider in measuring the people learnability on AR smart glasses”. In answering the aforementioned research question, Table 1 shows the summary of factors, identified from the related existing work on AR.

Table 1. Related Work of Learning Factors with AR Smart Devices

Author	Research Scope	Factors Used
Philipp A.Rauschnabel [4].	Augmented reality smart glasses: A study comprising of technology approval factors. An investigation strategy was directed for the evaluation of the empirical model.	Functional benefits, social norms, ease use, technology innovativeness, and adoption intentions.
Murat Akcayir, Gokce Akcayir [5].	The scope is to reveal the benefits, gains and encounters related with augmented reality in the field of education with the manuscript selection process.	Allowing learners to learn by doing, provide interaction opportunity, enable visualizations of invisible concepts events, abstract concepts, and help students to understand and enhancing learning.
Apurva Adapa,Fiona Fui-Hoon Nah, Richard H. Hall, Keng Siau, and Samuel N. Smith [6].	The scope is to check the factors Influencing the adoption of smart wearable devices through the qualitative approach and through in-depth interviews with students and staff at a Midwestern technological research university.	Efficiency, time-saving, tech novelty, information accessibility, professionalism.
Bacca, J., Baldiris, S, Fabregat, R. Graf, S, & Kinshuk. [7].	The scope is augmented reality trends in education using these approaches the qualitative-exploratory-case study, Qualitative-Exploratory-Pilot study and Qualitative-Exploratory-experience survey methodology is used.	Explaining the topic, facilitating interactions, just in time information, learning gain, better learning performance, exploration, lab experience.
Tien-Chi [8].	Animating eco-education to perceive, sense, and ascertain in the AR based surroundings with the Pre-test and post-test survey method through a model developed (EDALM) on Kolb's experiential learning theory.	Learning effectively.
Caudell, T. P, & Mizell, D. W [9].	The scope is on the social control front it provides valued insights towards the usage of the AR devices.	Focused advantages, enhanced focus, easy use, individual attention, thorough attitudes, and social norms.
Jorge Martín-Gutiérrez a, Peña Fabiani a, Wanda Benesova b, María Dolores Meneses c, Carlos E. Mora [10].	The Author discussed that the augmented reality in smart campus or education will pave path towards enhancement of learning absorption.	Learning absorption, easy understanding and no repeating while explaining also the improvement of learning motivation.

As presented in Table 1, researchers have dealt with AR Wearable devices in various scope of investigation based on their choice of factors. Factors that are suitable to be used as the learnability factors for our research have been chosen and discuss in the following section.

3. Initial Learnability Factors

Table 2 presented the choice of learnability factors that we found suitable for our initial investigation on measuring the level of learnability among the students in the use of AR Smart Glasses. There are seven (7) factors: enhanced the focus, enable visualizations of invisibility of concepts, promote imaginative thinking, enhanced imagination and information accessibility, better lab experience and interaction, easy and quick understanding and better learning and performance.

Table 2. The Chosen Learnability Factors and Description

Learnability Factors	Description
Information accessibility	Wearable smart glasses are like computers as they provide in time information access whenever and wherever needed [6].
Better lab experience	With the help of additional images video and audio in front of the user's eye this whole experience gives a better lab experience for the students and the users as it makes understanding easier and lab hours more fun.[7]. It also brings in additional functional benefits with AR smart glasses [4].
Better learning and performance	The usage of AR smart glasses gives just in time information which will give better learning performance [7]. AR smart glasses increases the learning performance of the user [4, 11, 12].
Enhanced focus	AR increases the focus of the users [9]. It also raises the level of engagement in users [13]. AR facilitates the learnability of the users in a smart campus.
Easy and quick understanding	AR smart glasses makes the understanding easy for the users [14]. As with the help of AR smart glasses one can see things happening so it becomes lot easier to understand by seeing it happen live.
Promote imaginative Thinking	AR smart glasses makes imaginary theories possible in Augmented reality through the smart glasses [5]. Therefore, with the help of AR smart glasses one can bring our imagination into reality which helps enhance the ability of imagination and thinking in a person.
Enable visualization of invisible concepts	AR allows imagining of unseen theories, events, and abstract concepts [5]. The exploration of imagination will make learning effective and quick to understand.

These factors have been chosen due to the connectivity of the factors to achieve the learning outcome as well to increase the learning performance in education.

4. Conclusion and Future Work

In the literature exploration, we found a gap linked to the effectiveness of AR devices specifically to AR smart glasses in increasing the student experience for learning in smart campus. Thus, this study reviewing the potential learnability factors in measuring the effectiveness of the AR smart glasses usage. We found seven (7) suitable learnability factors. These initial finding will be further investigated in the next work. A survey will be designed to a group of students in two (2) universities to understand the learnability influence with and without AR smart glasses based on the chosen learnability factors.

References

[1] Philipp A. Rauschnabel and Young K. Ro Augmented reality smart glasses: an investigation of technology acceptance drivers, Int. J. Technology Marketing, **11** (2) (2016).
[2] Muteeb Alahmari, Tomayess Issa, Theodora Issa, S. ZaungNau, Faculty awareness of the economic and environmental benefits of augmented reality for sustainability in Saudi Arabian universities in Journal of Cleaner Production, **226**, (July 2019) 259-269.
[3] Billinghurst, M. Augmented reality in education, new horizons for learning, (2002).
[4] Rauschnabel, P.A., Brem, A. and Ivens, B.S. 'Who will buy smart glasses? Empirical results of two pre-market-entry studies on the role of personality in individual awareness and intended adoption of Google Glass wearable's', Computers in Human Behaviour, **49** (8) (2015) 635–647.
[5] Murat Akcayir, Gokce Akcayir, Advantages, and challenges associated with augmented reality for education (2016).
[6] Apurva Adapa, Fiona Fui-Hoon Nah, Richard H. Hall, Keng Siau, and Samuel N. Smith Factors influencing the adoption of smart wearable devices, (2018).
[7] Bacca J., Baldiris, S., Fabregat, R., Graf, S., & Kinshuk. Augmented reality trends in education. (2014).
[8] Tien-Chi Huang, Chia-chen chen,Yu-Wen Chou, animating eco-education, Computers and education, (May 2016).
[9] Caudell, T. P, & Mizell, D. W. Augmented reality: An application of heads-up display technology to manual manufacturing processes, in proceedings of the twenty-fifth Hawaii international conference on system sciences. (1992).
[10] Jorge Martín-Gutiérrez a, Peña Fabiani a, Wanda Benesova b, María Dolores Meneses c, Carlos E. Mora Augmented reality to promote collaborative and autonomous learning in higher education.(Dec 2014).
[11] El Sayed, N. A., Zayed, H. H., & Sharawy, M. I. ARSC: augmented reality student card an augmented reality solution for the education field. Computers & education, **56** (4), 1045e1061. (2011).
[12] Huizenga, J. J., Admiraal, W. W., Akkerman, S. S., & Ten Dam, G. Mobile game-based learning in secondary education: Engagement, motivation, and learning in a mobile city game. Journal of Computer Assisted Learning, **25** (2009) 332–344.
[13] Liu, P.-H. E., & Tsai, M.-K. Using augmented-reality-based mobile learning material in EFL English composition: An exploratory case study. British journal of educational technology. (2013).
[14] Chiang et al. An Augmented Reality-based Mobile Learning System to Improve Students' Learning achievements and motivations in natural science inquiry activities (2014).
[15] M Abdul Majid, U Aickelin, PO Siebers, Human behaviour modelling: an investigation using traditional discrete event and combined discrete event and agent-based simulation, University of Nottingham (2011).
[16] BA JNR, MA Majid, A Romli, Sustaiable adoption and implementation in collaborative enterprise: A systematic Litraure review, Journal of Theoretical & Applied Information Technology 95 (9) (2017).

Acknowledgment
This research work has been funded and supported by Universiti Malaysia Pahang(UMP), Malaysia.

IOP Conf. Series: Materials Science and Engineering **958** (2020) 012006 doi:10.1088/1757-899X/958/1/012006

Application of Smart Materials in Civil Engineering and Architecture

SZOLOMICKI Jerzy[1, a *]

[1]Wroclaw University of Science and Technology, Wroclaw 50-370, Poland

Jerzy.Szolomicki@pwr.edu.pl

Abstract. Smart materials are the result of great advancements in material engineering and have different practical applications. This paper presents a review of selected smart materials applied in modern building structures from civil engineering and architecture perspective and determines what advantages they can bring.These benefits include building structure, interior design and façade solution. A very important causative factor resulting from the use of intelligent materials is obtaining a new aesthetic effect. This effect can be achieved through refined patterns, change of shape, presentation of interactive images.

1. Introduction

Stronger energy efficiency requirements and interactivity require modern building structures to be constructed using new materials and solutions. In order to design and build a sustainable structures which fulfil functionality and load bearing capacity under static and dynamic loads application of smart materials is necessary. For that reason, civil engineers and architects are seeking new materials and technologies which are going beyond standard way of thinking and design [1, 2].

Smart materials have the capability to respond to changes in their condition or the environment to which they are exposed, in a useful and controlled manner. The input that cause the change in smart material properties may be in the form of mechanical stress or strain, electrical or magnetic field or changes in temperature, pH, moisture and light. They are used in modern building structures and contribute in increasing load-bearing, control vibration, noise reduction and energy efficiency of structure.

The following types of smart materials can be distinguished [3]: piezoelectric, electroactive, photostrictive, thermostrictive, magnetostrictive, chemostrictive materials and fiber optic sensors. These materials are components of an intelligent structure that is an electronically reinforced with a physical framework. For example, piezoelectric materials can produce electricity under pressure or can change its dimensions under the influence of electricity. Piezoelectric controllable dampers generate friction force by pressing friction materials with actuators. There are other forms of intelligent materials, such as shape memory alloys, which can be applied as temperature sensors in ventilation systems or as actuators in detection devices. The properties of glass can be improved using nanotechnology. There are now self-cleaning systems with antimicrobial activity that can reduce pollution properties. Titanium dioxide represents a type of broad-spectrum bactericide with excellent biocompatibility which create an anti-adhesive coating.

IOP Conf. Series: Materials Science and Engineering **958** (2020) 012006 doi:10.1088/1757-899X/958/1/012006

2. Smart Materials in Civil Engineering

A group of intelligent materials that has an impact on the development of building structures and their safety is dynamically developed [4]. Smart concrete occupies an important place in this group [5]. By adding carbon fibres it is possible to add electrical impulses to the concrete structures, making the concrete able to have electrical resistance change in response to damage and deformation. This allows to detect small increases in internal stress. Signs of cracks or increase in stress can be monitored by the material itself. Due to shrinkage and other rheological phenomena which occur in reinforced concrete, cracks cannot be avoided. However, care must be taken to limit the propagation of such cracks. One possibility for minimizing cracking is through the use of fiber reinforcement such as steel, polyvinyl alcohol (PVA) or natural fibers [6] A recent and more reactive approach is the use of self-healing concrete: once a crack forms, it is autonomously repaired by the material itself. An example of intelligent material that can solve a problem is microbial concrete in which bacteria produce calcite (CaCO3), filling in the cracks, Fig. 1. In this process, bacteria produce urease which catalyzes the hydrolysis of urea into ammonium and carbonate.

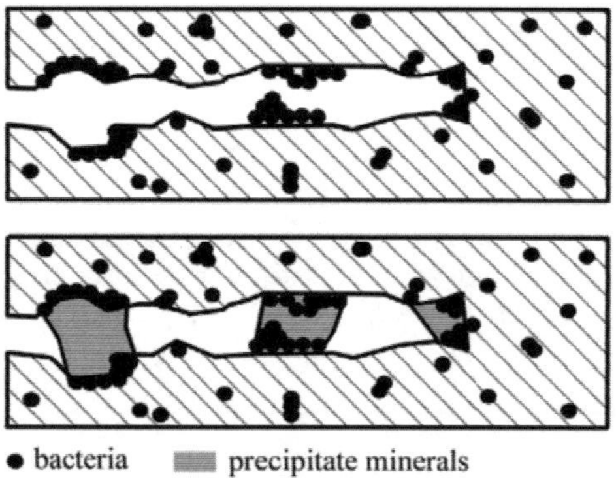

Figure 1. Schematic of crack healing by bacteria (developed by author on the basis of [7]

Another approach for reducing concrete cracking caused by thermal effects is incorporation of phase change materials (PCMs) in the concrete mix, Fig. 2. PCMs are a phase change materials that are able to absorb, accumulate and release a large amount of energy in the form of heat. In PCMs, heat is absorbed and released when the material changes its state from solid to liquid. Phase change materials can be divided in three categories: organic paraffin and non-paraffin compounds, inorganic (salt hydrates and metallics) and eutectic (composition of two or more components, each of which melts and freezes congruently forming a mixture of the component crystals during crystallization).

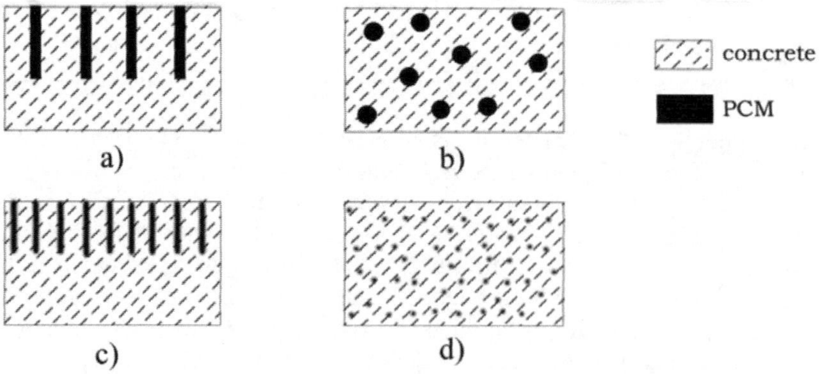

Figure 2. Different ways of incorporating PCM material into concrete: a) pipes filled with PCM; b) microcapsules with PCM; c) filling concrete surfaces by absorbing PCM; d) lightweight aggregate particles impregnated with PCM (developed by author on the basis of [6])

Another type of materials widely used in construction engineering are SMA (shape memory alloys) thermobimetals which react to changes in temperature by changing shape [8]. The reversible switching between the high temperature phase (austenite) and low temperature phase (martensite) and the associated changes in the electric, thermal and mechanical properties allow for passive control of structures. Due to its excellent deformation behavior and a very good resistance Ni-Ti, Cu-Zn-Al and Cu-Zn-Ni can be successfully used in damping of vibration [9]. Average residual inter-story drifts decreased with the use of SMA bracing system, Fig. 3. Moreover, tube connections elements made from SMA with one-way effect of shape memory could be used in construction of skeletal building elements such as lattice beams. Also, smart concrete beam are developed by taking advantage of the shape memory effect of SMA and characteristics of applying large forces on the resisting member in the transformation of SMA on heating [10]. In this case application of SMA wire allow concrete structure to recover from residual deformation by damaged earthquake after electrically heating the SMA wires. The effectiveness of self-restoration of SMA concrete beams is directly related to the increase in the initial pre-stress of the SMA wire.

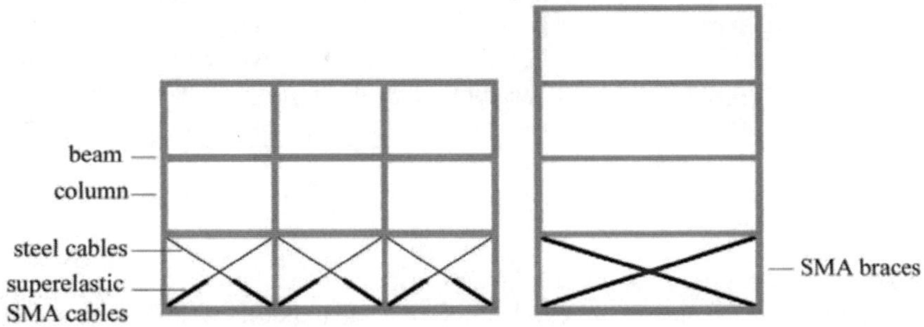

Figure 3. Schematic representation of SMA braces for steel building structure (figure by author)

Magnetorheological liquids (MR) are very important intelligent material used to ensure the safety of the structures. MR changes their viscosity, from a thick fluid to almost a solid, under the influence of a magnetic field. This dispersion system is composed of a carrier in the form of mineral or synthetic oil in which ferromagnetic suspension is dispersed. These materials are mainly used in vibration dampers of buildings located in seismic areas or subjected to strong wind action. The dampers are very often integral component of the main-load bearing structure, e.g. MR applied in Nihon Kagaku Miraikan building in Tokyo. An important application of magnetorheological dampers in building structures is also protection against displacement of bridge structures [11]. In structural solutions of suspension bridges they are used to damp cables of structure, Fig. 4. The damping parameters are regulated autonomously in the hydraulic damping element by way of electronic system. A sensor measures the natural frequency, sends data to computer which calculates the desired response force of the damper and sends its result back to the damper by way of electrical impulses [12]. This solution was used in Volgograd Bridge, which is one of the longest in Europe (Volgograd, Russia).

Figure 4. Scheme of vibration damping of the cable suspension bridge (figure by author)

IOP Conf. Series: Materials Science and Engineering **958** (2020) 012006 doi:10.1088/1757-899X/958/1/012006

3. Smart Materials in Architecture

A relatively large group of materials, more and more noticeable in architecture, are materials that change color or degree of transparency under the influence of specific factors. These are mainly glass materials [13, 14]. Some of them react to changes automatically, e.g. under the influence of light (photochromic glasses) or solar heat (thermotropic glasses). The most widely used are glasses with controlled variation of optical parameters, controlled by electricity. The most commonly used technologies are:

- *Electrochromic glass*

A special coating is placed between two glass panes filled with crystal particles in liquid form. The laminate consists of two layers of transparent conductors, electrolyte, electrochromic layer (active electrode) and ion storage (passive electrode). The change of the glass color to blue and its shading occurs due to the movement of ions between the electrodes under the influence of the applied voltage.

- *Suspended particle devices (SPD) glass*

The laminate consists of a thin film containing a suspension of elongated particles sandwiched between two sheets of transparent plastic or glass, Fig. 5a. In the case of absence of electrical voltage, suspended particles are arranged in random directions and tend to absorb light so that the glass plate is dark (gray or black). After applying voltage, suspended particles align their position and allow light to transmit. The potential of the applied voltage is controlled manually or automatically to precisely control the amount (and brightness) of light and heat passing through such glass, thus reducing the need for air conditioning in summer and heating in winter. An example of the use of this type of solution is the façade of Indiana University building in Bloomington (USA).

- *Polymer dispersed liquid crystal devices (PDLCD) glass*

The laminate consists of a thin film which contain a dispersion of liquid crystals in a suitably solidified or hardened polymer, Fig. 5b. As a result of a change in the state of polymer from liquid to solid, liquid crystals are separated from the polymer and in the form of droplets remain suspended in the polymer mass. In this system, a liquid crystal film is inserted between two layers of plastic or glass. Films with a thin layer of transparent conductive material bond the whole in the lamination process. In the initial state, the liquid crystals are arranged in a chaotic manner, which causes the glass to become opaque. After applying voltage crystals arrange in an orderly manner, and the glass becomes transparent. The transition from one state to another takes a very short time. An example of the use of this type of solution is the façade of Chanel Headquarter building in Tokyo (Japan).

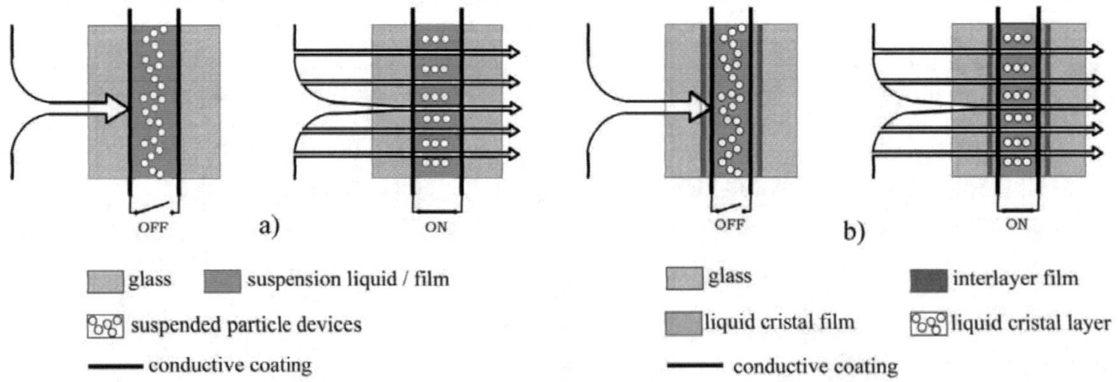

Figure 5. Scheme of two types of glass layers, before and after turning on the power: a) SPD glass layers; b) switchable glass with a spacer containing liquid crystals (developed by author on the basis of Hitachi Chemical Products)

- *Gas chromatic glass*

In gasochromic glazing technology, the color is determined by the active diluted hydrogen which flow through the void between the layers of the glass. This gas reacts with the active coating WO_3 causing its color. In the passive phase, the glazing is neutral and transparent. In the active phase it changes color to darkened navy blue. The advantage of this technology is short time of phase change. An example of the use of this type of solution is the façade of Frauenhofer Institute Building in Freiburg (Germany).

- *Micro-blinds*

Micro-blinds also control the amount of light passing through the glass when the appropriate voltage is applied. Micro blinds consist of many small and thin metal blinds installed on the glass pane. The metal layer is embedded in the magnetron sputtering process and the division into individual blind slats is carried out in the process of laser technique or lithography. The glass substrate contains a thin conductive layer (transparent conducting oxide TCO). Most often it is indium tin oxide (ITO), fluorine doped tin oxide FTO or doped zinc oxide. A thin insulator layer is embedded between a layer of thin metal blinds and a TCO layer. In the absence of voltage, the micro-blinds have rolled slats and light can pass through the structure.

Another type of activity using the features of intelligent materials can be light emission. After impulse operation, the particles go to a higher energy state, which lasts as long as the impulse works. Part of the energy is released in the form of visible light radiation which is not accompanied by heat radiation. Photoluminescent materials which react to light and electroluminescent which react to electric voltage are the most useful for architecture. They are used in the form of phosphorescent or fluorescent paint coatings, wallpapers or fabrics. LEDs are the best-known product from the group of electroluminescent materials. They are used to create media facades on the surface of which any moving images can be displayed. Initially, these were a grid structures applied to the facade surfaces. In the most modern solutions, the diodes are placed inside the glass pane. New generations of diodes based on polymer semiconductor (OLED) technology are also developed. An example of such a solution may be light-kinetic curtain wall façade in Hotel Habitat in Barcelona (Spain)

There are materials that are capable of producing electricity as a result of a specific stimuli, e.g. light, temperature, mechanical force or chemical environment. The most common material in this group are photovoltaic cells, obtaining electricity from solar energy. They use semiconductor technology, and their operation is reverse of LEDs. Individual cells are combined into modules and then into serial connection. New generation cells are produced as very thin films that can be bonded to the structure of a glass pane. The most modern products are semi-transparent glasses with PV cells, also colored in a very wide range of colors, and cells suitable for use on surfaces intended for pedestrians. The most effective solutions from an architectural point of view, e.g. semi-transparent cells, achieve lower efficiency.

Another very important example of intelligent solutions compatible with sustainable design is photocatalytic self-cleaning coating for building façade [15]. This solution incorporates nanometer-scale coating of titanium dioxide on the outer surface of façade. The photocatalytic activity of titanium dioxide is induced by the absorption of ultraviolet radiation. As a result of this process is initiated chemical oxidation of soiling deposits on exposed surfaces. The initial decomposition activated by the coating allows them to be easily washed under the influence of rain towards the façade of the building. Photocatalytic coatings revealed a great potential for air purification and easy cleaning properties. An example of the use of this type of solution is the Manuel Gea Gonzalez Hospital in Mexico.

4. Summary

The recently noticeable intensive development of intelligent materials paved the way for the development of building structures and changed the mindset of designers and specialists in the field of construction and architecture. Development of nanotechnology result in increasing number of new smart solutions. Many different functions that smart materials can achieve led to the composition of intelligent material systems that affect the behavior of structures. The potential of new smart materials technologies and material systems can change and improve the design and construction process. As they become more and more integrated with civil engineering and architecture applications they should be considered in the early stages of the design process. In the era of current technology, the development of sophisticated building structure is becoming extremely important because the creation of better structural solutions should be the driving force behind the development of technology, and not just a projection of what is available.

For example, application of SMA and MR is very effective in improving the response of structures to any extreme loading. Other concepts include self-repairing concrete, phase changing materials which enables create large surface area to improve heat transfer can be very efficient in reducing the cost of repairing and retrofitting of various structures.

Moreover, ecological solutions can be included self-cleaning buildings envelopes or thin polymer films mimicking the function of living skin which are adopted in a building envelope.

The use of intelligent glass allows savings by reducing the costs of heating, air conditioning and lighting, as well as avoiding the costs of installation and maintenance of automation and drives for traditional sun visors, such as blinds or awnings. Glass, which is not transparent in an inactive state, blocks almost all UV transmission through the glass, and thus reduces the fading of fabrics, paints and other products sensitive to UV radiation.

The overall conclusion is that material engineering tends to increase the ability of intelligent materials to shape the comfort of the internal environment, security and energy efficiency of the building, as well as its aesthetic and ecological features.

References

[1] J.Szolomicki, H. Golasz-Szolomicka, Technological Advances and Trends in Modern High-Rise Buildings, Buildings **9** 193 (2019) 1-32.
[2] M. Casini, Smart Buildings Advanced Materials and Nanotechnology to Improve Energy- Efficiency and Environmental Performance, Elsevier (2016) 1-355.
[3] A. Ritter, Smart Materials in Architecture, Interior Architecture and Design, Birkhäuser 2007.
[4] B.A. Mir, Smart Materials and Their Applications in Civil Engineering: An Overview, International Journal of Civil Engineering and Construction Science **4**(2) (2017) 11-20.
[5] A.M. Shelvay, Reinforced Concrete: Applicability of Smart Materials, Master Thesis, Clarkson University 1-46 (2011).
[6] B. Šavija, Smart Crack Control in Concrete through Use of Phase Change Materials (PCMs): A Review, Materials **11**(5), 2018.
[7] M. Rajczakowska, K.H. Cwirzen, H. Hedlund and A. Cwirzen, Autogenous Self-Healing: A Better Solution for Concrete, Journal of Materials in Civil Engineering **3** (9) (2019).
[8] L. Janke, C. Czaderski, M. Motavalli and J. Ruth, Applications of shape memory alloys in civil engineering structures – Overview, limits and new ideas, Materials and Structures **38** (2005) 578-592.
[9] S. Ghodke, R.S. Jangrid, Equivalent linear elastic-viscous model of shape memory alloy for isolated structures, Advances in Engineering Software **99** (2016) 1-8.
[10] A. Panigrahi, Application of Smart Materials in Structural Engineering, Available online: https://www.engineeringcivil.com/application-of-smart-material-in-structural-engineering.html, (Accessed on 10 April 2020)
[11] L. Scislo, Controllable Smart Materials in Civil Engineering, Technical Transactions, Issue 3, 1-B/2011 163-171.
[12] F. Gordaniejad, M. Saiidi, B.C. Hansen, O. Ericksen and F.K. Chang, Magneto-Rheological Fluid Dampers for Bridges, Journal of Intelligent Fluid Dampers for Control of Bridges **13**(2) (2002) 167-180.
[13] J. Marchwiński, Modern Technologies within Buildings Elevations. Smart Materials with Changeable Properties – Selected Examples, Przestrzeń, Ekonomia, Społeczeństwo 12/II (2017) 193-208.
[14] B.P. Jelle, T. Gao, The Utilization of Electrochromic Materials for Smart Window Applications in Energy-Efficient Buildings, TechConnect Briefs (2015) 226-229.
[15] A.A. Ćurĉić, Photocatalytic self-cleaning facades in architectural design, Architecture and Civil Engineering **16** (3) (2018) 425-436.

IOP Conf. Series: Materials Science and Engineering **958** (2020) 012007 doi:10.1088/1757-899X/958/1/012007

Application of Date Palm Ash to Improve Strength Characteristics of Sabkha Soil

Zaheer Abbas Kazmi

Department of Civil & Construction Engineering, College of Engineering, Imam Abdulrahman Bin Faisal University, Dammam, Saudi Arabia

zakazmi@iau.edu.sa

Abstract. Surface layer of earth's crust in Kingdom of Saudi Arabia is composed of loose sand with soluble salts, which is generally termed as Sabkha. Poor strength and durability characteristics of the Sabkha soil adversely challenge the safety and serviceability of infrastructure built on it. Therefore, improvement of its physical properties demands urgent attention of the engineering society. In this study, date palm ash (DPA), a local industrial waste, is utilized as an additive to improve strength characteristics of problematic Sabkha soil. A remarkable improvement in compaction characteristics and compressive strength is achieved. By investigating the effect of different proportions, the optimum quantity of additive is determined with respect to compressive strength.

1. Introduction

Since the prominent oil boom of the 1970s and demographic growth, the Kingdom of Saudi Arabia is rapidly developing infrastructure facilities which are mainly concentrated along the coastline. The soil in shallower depth of the region is mainly composed of loose sand with soluble salts, which is generally termed as Sabkha [1]. These salts occur as crusts on the surface or in the dissolved form in soil moisture [2]. Previous studies [3 -6] revealed that magnesium, calcium, sulphates, sodium, chlorides, and carbonates of calcium dominate the salt precipitation in the Middle Eastern desert region. Presence of crystalized salt makes the cementation bond very weak. The presence of high salt content, extreme environmental conditions, and fluctuation of ground water table lead to large changes in density, consistency, strength, and swelling and shrinkage characteristics of the soil [7 - 9]. All these changes make the soil highly variable and complex in nature and can drastically damage the foundations as well as other infrastructure above them, especially to lightweight structures [8, 10]. Expansive behavior of the salt bearing soil along with salt crystallization leads to differential settlements and cracking of the pavement structures [11].

The strength and durability characteristics of the problematic Sabkha soil need to be improved to the civil engineering requirements by mechanical and/or chemical stabilization methods, which will reduce permeability and compressibility of the soil mass and increase its shear strength. In chemical stabilization, cement, bituminous materials, lime, blast furnace slag, and fly ash are most commonly used binders [12, 13]. Due to its high strength, workability, durability and availability, Ordinary Portland Cement (OPC) has been most excessively used for ground improvement. Nevertheless, production and processing of cement consumes huge amount of energy and has a significant contribution in global greenhouse gas emissions which is threatening for the environment [14]. The increasing use of cement also affects surface runoff of water, vegetation growth, and quality of ground water. Furthermore, the irreversible hydration reaction of cement makes it almost impossible to retain the original untreated state of soil [15]. Furthermore, cement is also considered relatively expensive for its mass scale application of ground improvement. To address the shortcomings of OPC, different ash-based cementitious binders evolved in the field of geotechnical engineering. Some typical examples of these binder include oil palm shell (OPS), palm oil fuel ash (POFA), and biomass ash (BA). The ash based cementitious materials are deemed environment-friendly and they have comparable or even better characteristics than OPC [16, 17].

Date palm is one of the major agricultural products in arid and semi-arid regions. The Kingdom of Saudi Arabia has more than 23 million date palm trees which produce enormous mass of agricultural waste in the form of midribs, stems, leaves, fronds, and coir. It has been revealed in some studies that Saudi Arabia annually produces more than 200000 tons of date palm biomass. Usually, date palm wastes are disposed in landfills or burned in the farms which cause environmental pollution. Despite producing excessive amount of date palm ash (DPA), no significant attention has been paid to make its use in construction and ground improvement.

This article presents an improvement in strength and durability properties of the salt bearing desert sand (Sabkha) by the addition of date palm ash (DPA), a locally produced industrial waste. It has a dual benefit of improving ground condition and reducing the environmental waste. Since all other Gulf countries have similar trend with respect to soil compositions, behavior, and production of date palm ash, outcomes of this study could be equally applicable to the whole Gulf region.

2. Study Area and Methodology

Study Area. The soil employed in this study was sampled from King Fahad Suburb, a new residential city developed in the west of Dammam, Saudi Arabia (Fig. 1). The surface layer of the soil in the study area is loose sand with different types of salts in dissolved or crystalline form. Analyzing the soil investigation reports provided by the municipality, it has been observed that the ground water table is very shallow in the whole study area. The residential, commercial and community service structures built on the problematic soil of the study area experienced severe cracking and damages due to excessive total and differential settlement of the foundations. Currently, there is increasing pressure from the inhabitants for the municipality to provide lasting solution to protect their lives and properties. The date palm ash used in this study is obtained from a local palm recycling plant.

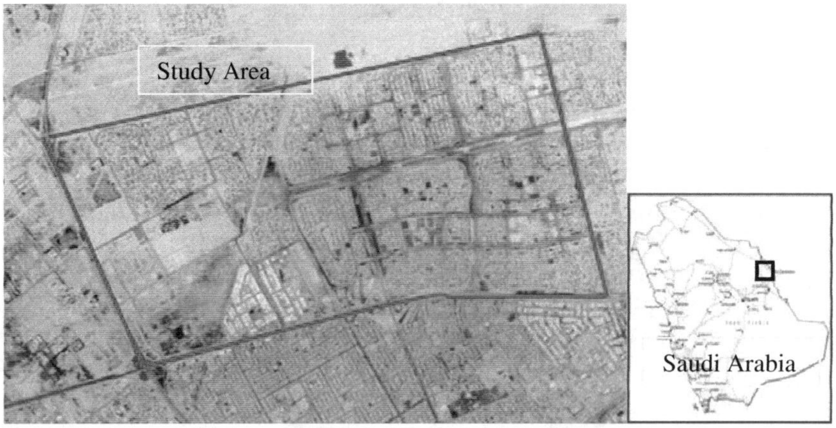

Figure 1. The Study Area (King Fahad Suburb, Eastern Province, Saudi Arabia)

Methodology. The DPA is mixed in different proportions of 2.5%, 5%, 7.5% and 10% by wight to the oven dried soil. The required amount of DPA is thoroughly mixed in the dry soil passing sieve number 4.

For Unconfined Compression Test (UCT), cubical specimens (5cm x 5cm x 5cm) were prepared for all the proportions of DPA. The specimens were properly cured for seven days before testing. The standard ELE equipment is used to perform sieve analysis, standard proctor compaction, and unconfined compression tests in order to determine grading, compaction, and strength attributes of the collected soil samples. ASTM standard procedures were followed in performing all the experiments.

3. Results and Discussion

Gradation. The sieve analysis tests were performed on the collected samples in accordance with ASTM D6913 to obtain soil gradation. It is very clear from the gradation curve (Fig. 2) that the surface layer of the soil in the study area is poorly graded, dominated by the percentage of sand. The percentage of gravels, sand and fines is found to be 6.5%, 90.5%, and 3.0%, respectively. The soil is classified as "*A-3(0)*" according to American Association of State Highway Transportation Officials (AASHTO), while "*SP (poorly graded sand)*" according to Unified Soil Classification System (USCS).

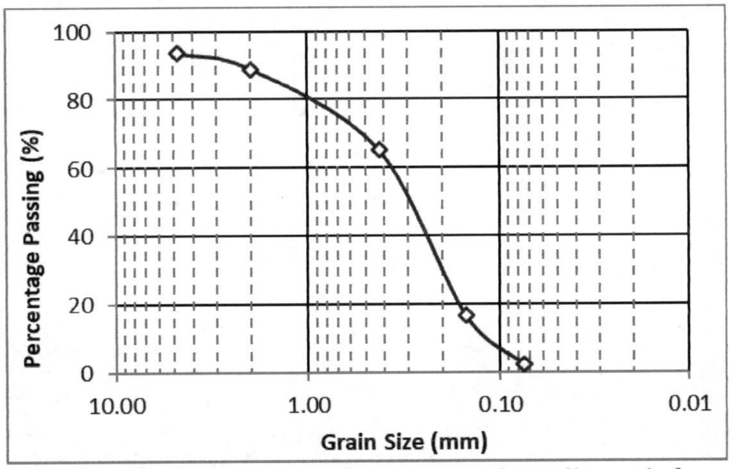

Figure 2. Grain size distribution curve of a representative soil sample from the study area

Compaction Characteristics. Following ASTM D558, the compaction characteristics of the collected soil samples were determined in their natural condition as well as with different proportions of DPA. The tests were performed initially by adding 7% of water to the oven dried soil samples, followed by an addition of 4% each time until we get at least two (02) points to ensure decline of compaction curve.

Figure 3 represents compaction curves along with variation of optimum moisture content (OMC) and maximum dry density for different proportions of DPA. It is evident from both the figures that the maximum dry density is continuously increasing with the increased proportion of DPA. The fine DPA grains are filling the voids between the sand particles, giving a higher maximum dry density. The optimum moisture content is also increasing with the increased proportion of DPA. It is well in line with the literature studies that a soil with more fine particles will have higher OMC. This is due to the fact that more water is required to make a lubricating film around increased number of particles.

IOP Conf. Series: Materials Science and Engineering **958** (2020) 012007 doi:10.1088/1757-899X/958/1/012007

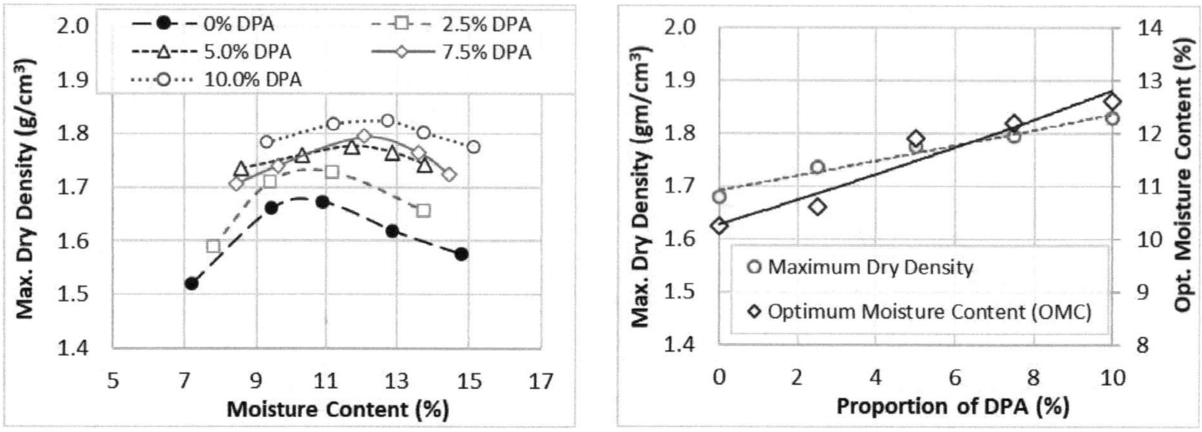

Figure 3. Compaction curves for the soil samples with different proportions of DPA (left), and variation of maximum dry density and OMC with different proportions of DPA (right)

Unconfined Compressive Strength. To investigate the cementitious properties of DPA, the soil samples with all the adopted proportions of the additive were tested in Unconfined Compression testing machine in accordance with ASTM D2166. Figure 4 represents compressive strength of all the tested samples. The results showed a remarkable increase in compressive strength up to 7.5% of DPA and a decline after that. The decline of compressive strength could be attributed to the domination of fine content in the sample over cementitious properties of DPA.

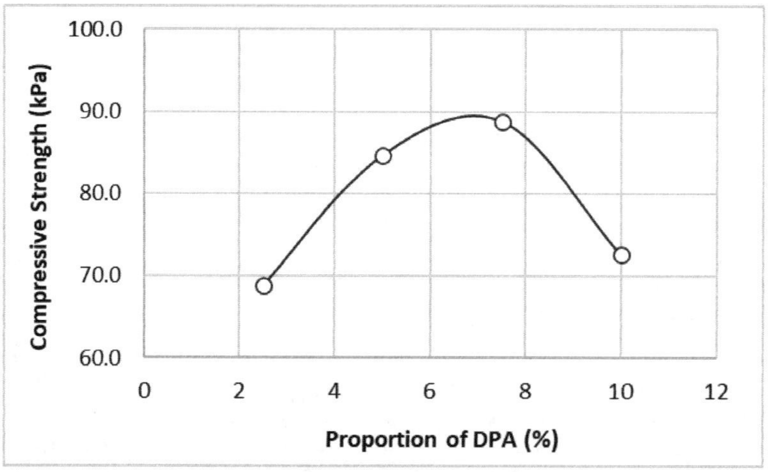

Figure 4. Effect of different proportions of DPA on Unconfined Compressive Strength of Sabkha soil

4. Conclusions

Surface layer of earth's crust in Kingdom of Saudi Arabia is composed of loose sand with soluble salts, which is generally termed as Sabkha. This Sabkha soil has poor strength and durability characteristics which adversely effect the infrastructure built on it. Therefore, it needs to be improved to the civil engineering requirements by mechanical and/or chemical stabilization methods.

In this study, a remarkable improvement in compaction characteristics and compressive strength of the Sabkha soil has been achieved by adding different proportions of date palm ash (DPA), a locally available industrial waste. Investigation has proved 7.5 % of the date palm ash as the optimum proportion of additive giving maximum compressive strength.

References

[1] A.G. Bauman, D.A. Feary, S.F. Heron, M.S. Pratchett, J.A. Burt, Multiple Environmental Factors Influence the Spatial Distribution and Structure of Reef Communities in the North-eastern Arabian Peninsula. Marine Pollution Bulletin, **72** (2013), 302 – 312.

[2] A.S. Stipho, Soil conditions and foundation problems in the desert region of the Middle East, In Proceedings of the International Conference on Case Histories in Geotechnical Engineering. Missouri University of Science and Technology, Missouri, United States, 6-7 May, 1984.

[3] S. Al Sayari, J. Zotl, Quaternary period in Saudi Arabia. Spinger-Verlag wien: New York, U.S.A., 1978.

[4] H.J. Barth, B. Böer, Sabkha Ecosystems: Volume I: The Arabian Peninsula and Adjacent Countries, Vol. 1. Kluwer Academic Publishers: Dordrecht, Netherlands, 2002.

[5] A.S. Stipho, Effect of Salt Concentration on some of the engineering properties of soil in Saudi Arabia. In Proceedings of Symposium on Geotechnical Problems, King Saud University, Riyadh, Saudi Arabia, May 1981.

[6] A.S. Stipho, Inherent Foundation Problems in Salina Soil. First Saudi Engineering Conference, Jeddah, Saudi Arabia, 1983.

[7] D. Hossain, K.M. Ali, Shear strength and consolidation characteristics of Obhor Sabkha, Saudi Arabia. Quarterly Journal of Engineering Geology and Hydrology, **21**(1988), 347–359.

[8] O.S.B. Al-Amoudi, Studies on Soil-Foundation interaction in the Sabkha Environment of Eastern Province of Saudi Arabia. PhD Dissertation, King Fahd University of Petroleum and Minerals, Dhahran, Saudi Arabia, 1992.

[9] A.N. James, A.L. Little, Geotechnical aspects of sabkha at Jubail, Saudi Arabia. Quarterly Journal of Engineering Geology and Hydrology, **27** (1994), 83 – 121.

[10] W. Akili, Foundations over salt-encrusted flats (sabkha): profiles, properties and design guidelines. In Proceedings of the 5th international conference on case histories in geotechnical engineering, New York, U.S., Paper **1** (43) 19, 2004.

[11] A.A. Bubshait, Quality of Pavement Construction in Saudi Arabia, Practice Periodical on Structural Design and Construction, **6** (2001), 129 – 136.

[12] A. Arulrajah, S. Horpibulsuk, F. Maghool, Recycled construction and demolition materials in pavement and footpath bases. The 6th International Symposium on Rural Roads, Bangkok, Thailand, 2016

[13] A. Kampala, S. Horpibulsuk, N. Prongmanee, A. Chinkulkijniwat, Influence of wet-dry cycles on compressive strength of calcium carbine residue-fly ash stabilized clay. J. Mater. Civ. Eng., **26** (2014), 633-643.

[14] H. Chen, Time-dependent reliability analysis of coastal defences subjected to changing environments, Structural Monitoring and Maintenance, **2**(2015), 49-64.

[15] I. Chang, J. Im, G.C. Cho, Introduction of microbial biopolymers in soil treatment for future environmentally-friendly and sustainable geotechnical engineering. Sustainability (Switzerland), **8**(2016), 251.

[16] D. Mujah, Compressive strength and chloride resistance of grout containing ground palm oil fuel ash, J. Cleaner Product. **112**(2016), 712–722.

[17] N.G. Ozerkan, B. Ahsan, S. Mansour, S.R. Iyengar, Mechanical performance and durability of treated palm fiber reinforced mortars. Int. J. Sustain. Built Environ. **2**(2013), 131–142.

IOP Conf. Series: Materials Science and Engineering **958** (2020) 012008 doi:10.1088/1757-899X/958/1/012008

Mechanochemical Production of Urea-citric Acid Copolymer

Eval O. Miranda Jr.[1,a *], Victor H. R. Silva[1,b], Mirtânia A. Leão[2,c], Elaine C. M. Cabral-Albuquerque[1,d], Silvio Cunha[3,e], and Rosana L. Fialho[1,f]

[1]Universidade Federal da Bahia/Programa de Pós-Graduação em Engenharia Industrial, Escola Politécnica, Rua Aristides Novis, n° 02, Federação, CEP: 40210-360, Salvador, BA, Brazil.
[2]Instituto Federal de Educação, Ciência e Tecnologia da Bahia/Programa de Pós-Graduação em Engenharia de Materiais, Rua Emídio dos Santos, S/N°, Barbalho, CEP: 40301-015, Salvador, BA
[3]Universidade Federal da Bahia/ Programa de Pós-Graduação em Química, Instituto de Química, Rua Barão de Jeremoabo, n° 147, Ondina, CEP: 40170-115, Salvador, BA, Brazil.

eval_oliveira@hotmail.com

Abstract. Increasing non-biodegradable plastic consumption is a great factor that contributes to worldwide pollution and global warming. Mechanochemistry provides a clean, solvent-less, route for chemical reactions such as polymerization based on the absorption of mechanical energy. Urea polymers have several applications in industrial and medical area, such as paints, resins, plastic, wood glue, separating aliphatic hydrocarbons, coating, and pharmaceutical drugs. In this study, a copolymer of urea and citric acid was synthesized using a planetary ball mill as grinder with fixed reaction time and rotational speed, varying the number and material of the planetary ball mill sphere to find the effects of those two conditions in the yield percentage of the copolymer. The purified polymer was characterized by Infrared spectroscopy (FTIR) and Nuclear Magnetic Resonance – NMR. Results confirmed that polymeric material is formed by the reaction between citric acid and urea without any catalyzer using mechanochemistry. NMR results showed that the reaction occurred between urea and the acid groups at the ends of the chain, while the central COOH remained intact.

1. **Introduction**

Generally, polymerization reactions are carried out in reactor in the presence of organic solvent medium. That fact creates many problems such as solvent loss, need to recover and recycle solvents, and contamination of solvent in the polymeric material. Different polymerization techniques were developed to obtain polymers without the use of organic solvent, like supercritical polymerization and mechanochemical reaction. The mechanochemistry's goal is to transform energy from mechanical forces to promote interaction between chemical structures. That energy can be provided to the different set of mechanical actions such as impact, compression, shearing and grinding [1]. So, a mechanical activation technique allows to perform the polymeric synthesis without organic solvents [2]. Besides, mechanochemical reactions provide excellent control of the reaction stoichiometry [3] and higher reaction rates due to absence of solvents [4]. In this context, the following work aims to produce urea and citric acid copolymers via mechanochemical synthesis and characterize the polymeric material by infrared spectroscopy and Nuclear Magnetic Resonance – NMR.

2. Experimental

Materials

All reagents and solvents were purchased as analytical grades from their respective suppliers and used as received. Urea ((NH2)2CO), M.M. = 60,07 g/mol, 99,5% minimum purity, supplied in the form of small granules by Petróleo Brasileiro S.A. Citric acid (C6H8O7) M.M. = 192,124 g/mol, purity > 99% supplied as small grains by Merck Comércio Importação LTDA. All the reactants were previously analyzed via FTIR (Fourier-transform infrared spectroscopy) to assure their purity.

Methods

Mechanochemistry synthesis. For the experiments, Adventurer® Precision Electronic Balance, AR2140BR - OHAUS with precision of 0,001g and a Retsch model PM-100 Planetary Ball Mill were used. Briefly, urea and citric acid were weighted, approximately 10 g of total mass, following a molar ratio of 1:1 [5]. After that, the monomers were transferred to stainless steel, agate and tungsten carbide vessels (50 cm3), which were loaded with balls 10 mm in diameter. The system was milled separately in the mill without solvent, as suggested by Landim et al. (2017) [6]. Reaction conditions were fixed in 550 rpm rotational speed, 60 minutes to polymerization reaction (with 1 minute pause every 5 minutes), applying direction reversal on each break as suggested by Landim (2018) [7].

Product purification (Filtration and Drying). The product purification (after synthesis) was performed by vacuum filtration using ethanol as wash solvent. Ethanolic solution was dried in a rotary evaporator (IKA HB10) with bath previously heated until 50 °C under stirring (80 rpm) in order to determine the residual content. After purification, the resulting polymer was dried in an oven (Quimis' 0317M-42) at 65 °C for 6 hours. After that, polymeric material was led to a silica-filled glass vacuum desiccator for 30 minutes until room temperature was achieved.

Reaction Yield. The yield of the copolymer synthesis was calculated through Equation 1, where n is the yield in percentage, m1 is the copolymer weight and m2 is initial reactant mass.

$$n\% = \frac{m1\ (100)}{\mathrm{m2}},$$

Infrared spectroscopy (FTIR). Infrared spectroscopy analysis was performed by Shimadzu FT-IR IRAffinity-1 equipped in an ATR device with a Germanium-coated KBr beam splitter and a DLATGS detector in the wavelength region of 4000-400 cm-1 with a precision of 4 cm-1 to observe the polycondensation/esterification process and to estimate the extent of the reaction. An average of 128 scans was recorded.

3. Results

Variables X Polymer Yield

In this study, fixed values of time (1h) and rotation (550 rpm) were adopted. The studied variables were the number of spheres and the type of material of the spheres, and the response variable was the reaction yield. 10 reaction runs were carried out in duplicate (Table 1). The highest yields were obtained with 02 spheres. In this way, the number of spheres was fixed at 02, and the type of material used was varied, to evaluate the influence of the composition of the material on the reaction yield.

Table 1. Effect of the number of spheres on the mechanosynthesis of the UAC copolymer

Run	Spheres	Yield (%)
1	1	77
2	1	74
3	2	82
4	2	79
5	4	71
6	4	68
7	6	77
8	6	74
9	2	80
10	2	78

Table 2. Influence of material composition on polymer yield

Constitution of the reactor and spheres	Yield (%)
Stainless Steel	82
Tungsten Carbide	70
Agate	52

Three different materials were evaluated, stainless steel, tungsten carbide and agate. And the best yields were obtained with stainless steel. Table 2 shows the yield values obtained by varying the type of material used. As the density of stainless steel is higher, it was expected a higher yield using this material, which was confirmed by the results presented in Table 2.

Concerning the IR analyses, it was expected to occur peaks in the regions of 3480 to 3200 cm-1 and 1720 to 1650 cm-1 for the urea, and 1750 to 1600 cm-1 for acid citrus [5]. This was confirmed in the analyzes performed, which are shown in Figures 1. In this case, the N - H bands in the region of 3437 cm-1 and 3360 cm-1 were observed, and the presence of energy peaks in 1670 cm-1 and 1634 cm-1 which is a band characterized by the presence of urea carbonyl. As for citric acid, the characteristic peak was observed in the 1693 cm-1 region that characterizes a carboxylic acid.

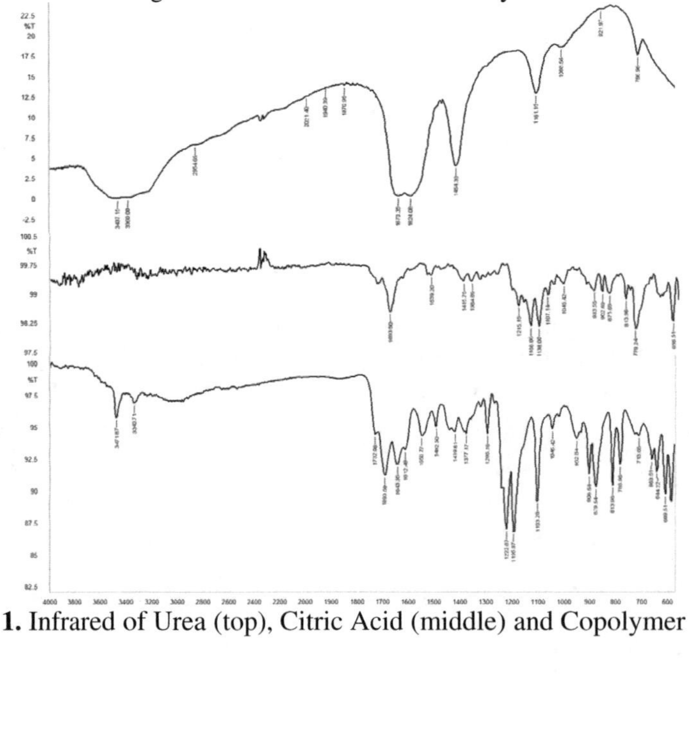

Figure 1. Infrared of Urea (top), Citric Acid (middle) and Copolymer (botton)

The formation of amides in the region of 3841 cm-1 and 3344 cm-1 was verified for the polymer. The 1643 cm-1 band suggests the presence of the carbonyl amide group (-NH-C = O) formed through reaction between citric acid and ureia. The formation of amides attest that mechanochemistry is efficient in transforming mechanical energy into chemical energy necessary for the condesation reaction.

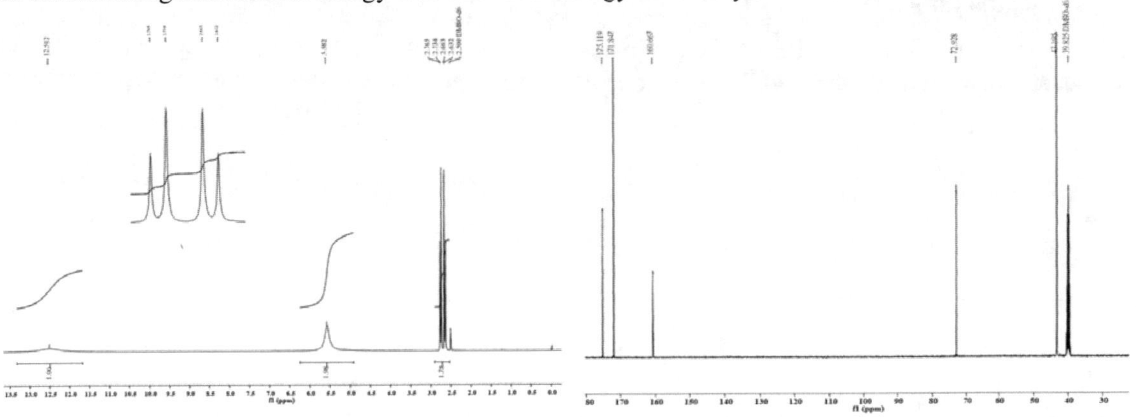

Figure 2. ^{1}H (left) and ^{13}C (right) Nuclear Magnetic Resonance Spectra of copolymer in DMSO-d6

The hydrogen and carbon nuclear magnetic resonance (NMR) spectra of the copolymer obtained are shown in Figures 2. In the 13C NMR the presence of five carbon atoms is observed, three of which refer to carbonyls (160.6 to 175.1 ppm) as well as the presence of two aliphatic carbons, one carbinolic at 72.9 ppm and the other at 39.8 ppm, corresponding to the citric acid portion. The two spectra correspond to the symetrical unit of chemical formula below, whose signals reveal that the reaction occurred between urea and the acid groups at the ends of the chain. The carboxylic acid unit of the central part of the chain is also confirmed in the 1H NMR spectrum where the presence of a hydrogen at 12.5 ppm is observed for one unit of hydrogen.

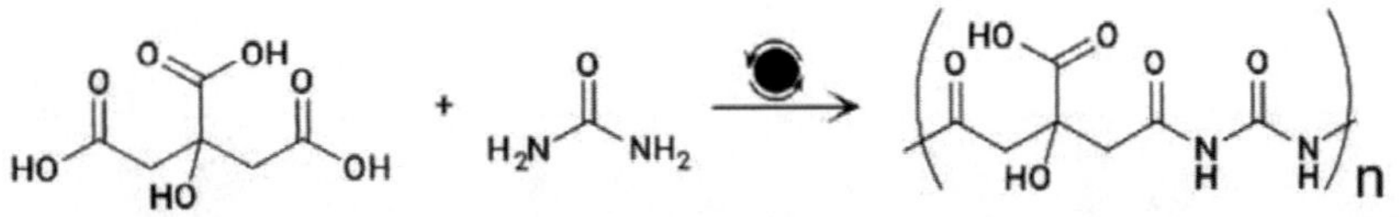

Based on the results, it is possible to claim that the polymer synthesis yield was 82% and the polymeric material is soluble in water but insoluble in ethanol. Thus, it was confirmed that the reactor is efficient in promoting polymeric synthesis, generating the polymer claimed by an easy-to-handle and environmental-friendly route.

References

[1] Tan, D.; Garcia, F. Main group mechanochemistry: From curiosity to established protocols. Chem. Soc. Rev. 2019, **48** 2274-2292.
[2] Chaves, C.M.M. Novas Metodologias em química sustentável. Universidade de Coimbra. 2015.
[3] Schmidt-Naake, G.; Frendel, A.; Drache, M.; Janke, G. Chemical Engineering & Technology, **24** 889 2001.
[4] Wang, G.W., Komatsu, K., Murata, Y., Shiro, M., 1997. Synthesis and X-ray structureof dumb-bell-shaped C120. Nature **387** 583.
[5] Paleckiene, R.; Sviklas, A.; Slinksiene, R. Reaction of urea with citric acid. Russian journal of applied chemistry, **78** (10) 1651-1655, 2005.
[6] Landim, Lucas B. et al. Synthesis and characterization of copolymers of urea–succinic acid–ethylene glycol and copolymers of urea–succinic acid–glycerol. Polymer Engineering & Science, **58** (9) 1575-1582, 2017.

IOP Conf. Series: Materials Science and Engineering **958** (2020) 012008 doi:10.1088/1757-899X/958/1/012008

[7] Landim, Lucas B. et al. Synthesis and characterization of copolymers of urea–succinic acid–ethylene glycol and copolymers of urea–succinic acid–glycerol. Polymer Engineering & Science, **58** (9) 1575-1582, 2018.

Acknowledgments

The authors wish to thank CAPES for their financial support. GPSQ/UFBA for the availability of the mechanochemical reactor and IQ/UFG for the RMN test.

IOP Conf. Series: Materials Science and Engineering **958** (2020) 012009 doi:10.1088/1757-899X/958/1/012009

Evaluation of the energy driving performance of a cooling system assembled with a Peltier module operated in hot climates at different electrical currents

David J. Jimenez-Rodriguez[1, a], Carlos A. Sanchez-Ramos[1, b], Diana Palma-Ramirez[2, c], Juan Carlos Paredes-Rojas[2, d], Abelardo I. Flores-Vela [2, e] and Facundo J. Marquez-Rocha [1, f] *

[1]Centro Mexicano para la Producción más Limpia-UT, Instituto Politécnico Nacional, TBC, Carretera Reforma - Dos Bocas, Km. 17+920, Cunduacán, Tabasco, C.P. 86126, México.
[2]Centro Mexicano para la Producción más Limpia, Instituto Politécnico Nacional, Av. Acueducto s/n, la Laguna Ticomán, C.P. 07340, México City, México.

fmarquez@ipn.mx

Abstract. Peltier devices have two sides with two different temperatures when an electrical current is applied through it. In the hot side, heat is dissipated and in the cool side, heat is absorbed, requiring additional electrical current to decrease heat absorption, primarily in hot climates. In this work, a block model was design with a Simulink tool of Matlab software to analyze thermal performance of a Peltier module in hot-imitate climate at different electric current levels. Through this model, it would be possible to evaluate the operating conditions of different control systems for Peltier applications, such as energy optimization in human thermal comfort devises, for sports, mobility, and medical care applications.

1. Introduction

Thermoelectricity was discovered in the early 1800s by a Thomas Seebeck, how applied a voltage through a junction of two different materials generating different temperatures. A Peltier module consists of the union of several pairs of semiconductors either p-type thermocouples (free holes) or n-type (free electrons). Semiconductors are thermally connected in parallel to reduce the thermal resistance, and electrically connected in series [1] to increase the operating voltage; these semiconductors are assemble between two ceramic plates [2, 3], as showed in Figure 1. A Peltier device works as a cooler-heater when it operates as a power generator or in Seebeck thermoelectric effect.

IOP Conf. Series: Materials Science and Engineering **958** (2020) 012009 doi:10.1088/1757-899X/958/1/012009

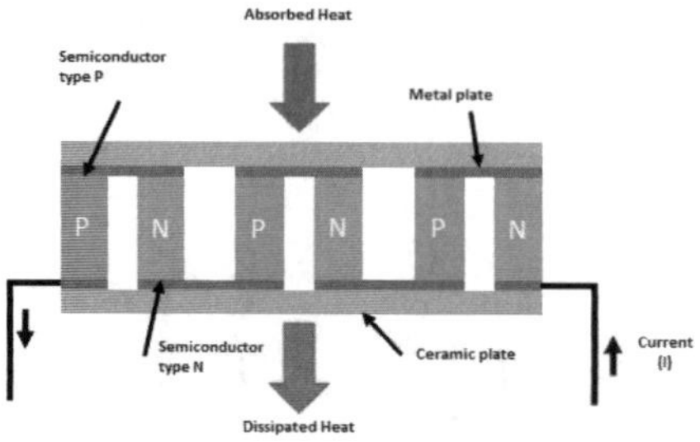

Figure 1. Peltier module components.

Peltier devices are solid-state energy converters, which operate without any mechanical or movement elements. These devices have some advantages compared to other equipment of similar applications such as low noise, compactness, reliability, long service life and stability [4,5]. Moreover, the Peltier modules are light, they do not use any refrigerant fluid and the cycle can easily switch between cooling and heating [6]. The main disadvantage of Peltier modules is their low-level efficiency or coefficient of performance (COP). Therefore, applications have been limited to scientific, spatial and medical equipment, where efficiency is not the most important factor [6]; however, this technology is now being used in cooling or warming on human comfort applications, such as, portable refrigerators [7], electronic circuit coolers, vehicle seat coolers and heaters [8].

In order to simulate the behavior and analyze the energy efficiency of the Peltier modules, some mathematical models for cooling-heating and power generation have been developed. The modeling and simulations allow the analysis, design and optimization of the Peltier modules with aim to obtain the energy performance of one or several modules for a specific application. Huan Liang Tsai [5] use the Matlab-Simulink software to construct a model of a Peltier module using block design, where the interaction of the input data is carried out through an interface, which allows to run a simulation with few data as well as exiting data, such as voltage and current requirements at certain difference of temperatures between the hot and cold sides of the Peltier.

In this work, a block diagram through Matlab-Simulink platform and mathematical equations of Peltier thermo electrical effect serve to analyze the thermal behavior of Peltier heat movement in base one experimental data and identification of heat sinks. The simulation considers environmental variables such as climate temperature and power supply. In the simulation process temperature data from both, the cold face (heat absorption), and hot face (heat dissipation), heat dissipated, heat absorbed, electrical power and the coefficient of performance (COP) can be obtained. The critical parameters for the operation of thermoelectric modules were identified and the influence of ambient temperature on the operation of the Peltier module obtained with the block model with a friendly-user interface for input data which allow the evaluation of the thermal and electrical behavior of Peltier module systems.

2. Operation Principles of thermoelectric modules

In cooling operations, the heat supplied to the hot side should be dissipate by a sink with the lowest possible thermal resistance, (Figure 2).That is, if the temperature of the hot side increase and the heat absorbed is lower, the temperature of the cold side increase, consequently, the COP decrease, and the electrical power required for module operation increase. When the temperature difference is constant at a fixed electric power; that is, if the temperature of hot side is increased, the temperature of the cold side increase as well [9].

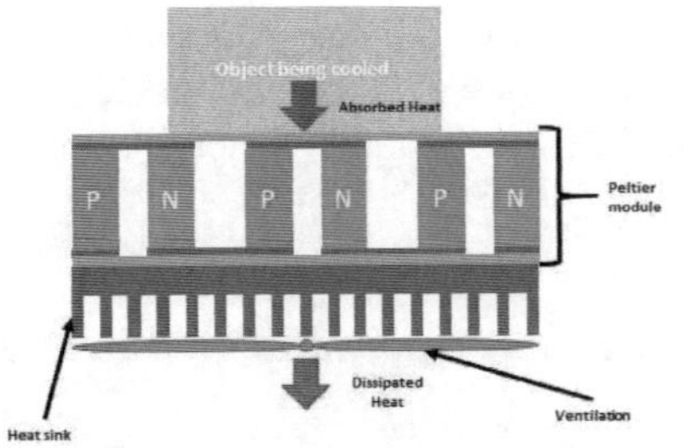

Figure 2. Cooling system using Peltier module.

Peltier modules can be considered as thermal systems because they involve energy transfer in two forms, one of heat dissipation on the hot side and the other of heat absorption on the cold side of the module. If the thermal system is analyzed in terms of thermal resistance and capacitance [10], Thermal resistance R_t (° C / W or °K / W) is defined as the ratio of the temperature gradient and the heat flux per unit of time, transported through a thermal conductor [10].

$$\mathbf{R}_t = \frac{\Delta T\,(t)}{Q\,(t)}. \quad (1)$$

The following expression represents the differential equation of thermal capacity [10].

$$C \cdot \frac{dT}{dt} = Q(t) \Rightarrow T(t) = T(0) + \frac{1}{C} \cdot \int_0^t Q(t). \quad (2)$$

3. Model building

A Peltier module has two sides, heat sinks for both sides of the module; this is a key factor to consider for the development of a model, cold side (heat absorption) and hot side (heat dissipation). The heat dissipation in this model is carried out by an electric fan. To fix properly, the use of thermal paste between the heat sink and the Peltier device is required. Heat sink and fan devices were assembled to generate the following data: Thermal resistances on the cold and hot sides (Rc and Rh), thermal capacities of the cold and hot sides (Cc and Ch) and thermal resistance of the paste used between the heat sink and the Peltier device (Rs). The thermal resistances and capacities of the hot and cold faces of the module during the absorption and dissipation of heat towards the environment are represented in Figure 3.

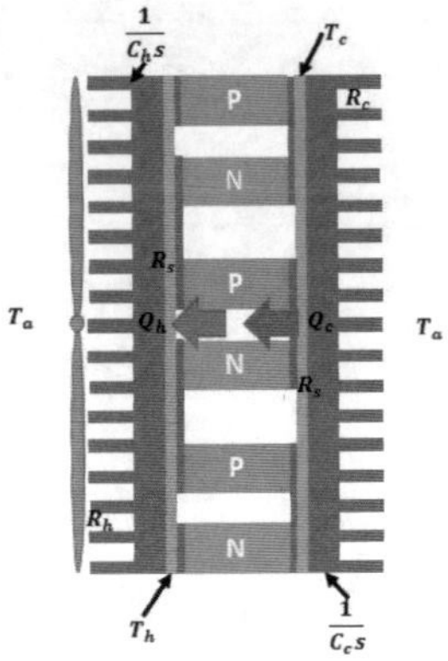

Figure 3. Scheme of the Peltier module, interpretative localization of resistances and thermal capacitances which intervene in the absorption and dissipation of heat along the system.

IOP Conf. Series: Materials Science and Engineering **958** (2020) 012009 doi:10.1088/1757-899X/958/1/012009

Considering that the energy capacity of the heat-sinks and if each heat-sink is isothermal, these can be considered as capacitive elements. By solving the previous equation for both sides of the module and rebuilt to Laplace transform, the following transfer functions were obtained:

$$\frac{T_h - T_a}{Q_h} = \frac{R_h + R_s}{1 + C_h.s.(R_h + R_s)}. \tag{3}$$

$$\frac{T_a - T_c}{Q_c} = \frac{R_c + R_s}{1 + C_c.s.(R_c + R_s)}. \tag{4}$$

T_a (K) is the ambient temperature, R_h (K/W) and R_c (K/W) are the thermal resistance of the hot and cold side of the Peltier module and the air, C_h and $C_c$is the thermal capacity of both sidesand s is the Laplace variable used to express the equations as a function of transfer.

It should be noted that the thermal capacity depends of the size of the heat-sink and the thermal resistance depends on the following factors:

- The thermal contact between the Peltier module and the heat-sink.
- The heat-sink model
- The voltage with which the fans are fed, that is the speed at which the air through the fan moves.

Once defined the principles than run the thermal system and considering the effects and mathematical expressions that lead the behavior of the Peltier module, a block diagram model was constructed to represent the dynamics of the Peltier module in the Simulink Matlab software. The Figure 4 shows the block model obtained from the Peltier module while operating in cooling mode, where the red color represents the hot side and the blue color represents the cold side of the Peltier module. The Table 1 shows the input data required by the model to evaluate the different variables involved in the operation of the Peltier module.

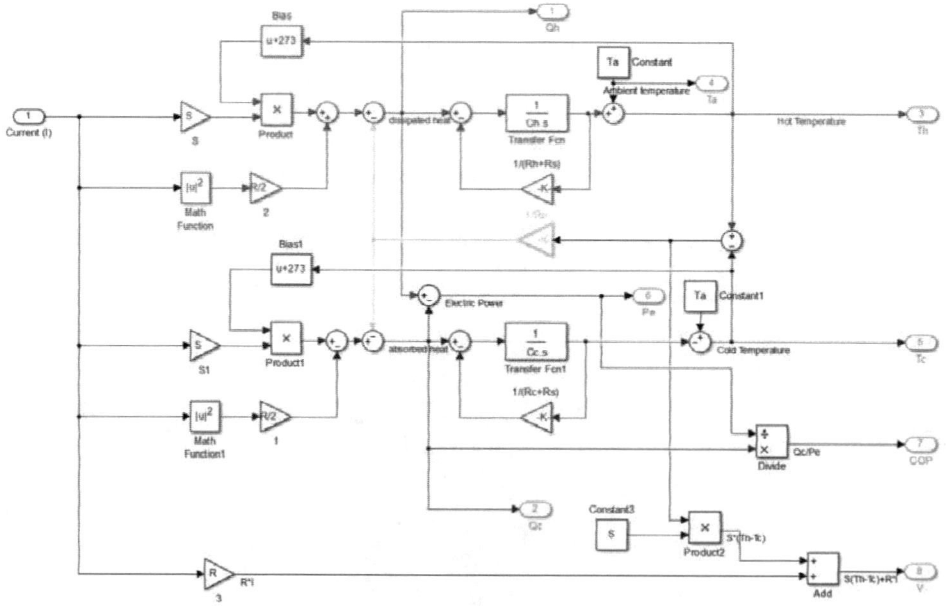

Figure 4. Diagram of the Peltier module.

Table 1. Input parameters.

Parameters	Unit
Seebeck Coefficient	V/K
Electric Resistance	Ω
Thermal Resistance Module	k/W
Thermal Resistance Cold Side	k/W
Thermal Resistance Hot Side	k/W
Thermal Resistance Grease	k/W
Thermal Capacity Hot Side	J/K
Thermal Capacity Cold Side	J/K
Ambient Temperature	(°C)

4. Model validation

Some graphs were generated in order to compare the behavior of the technical data of the Peltier module and the behavior obtained according to the simulations of the developed model.

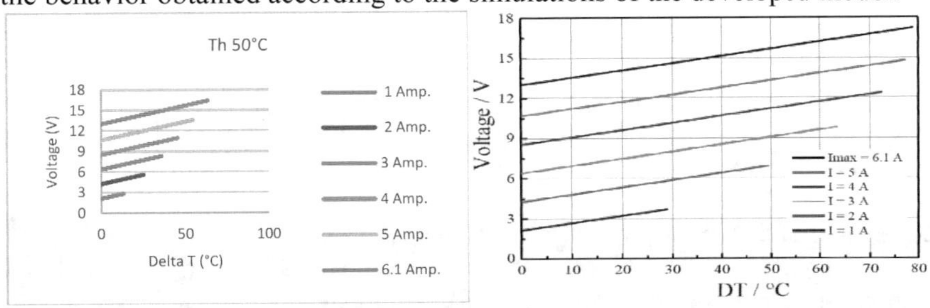

Figure 5. Behavior of the voltage supplied to the Peltier module, related to difference in temperature between the hot and cold side, and the electric current.

Figure 5, shows that the temperature differences obtained from the simulations (left) were smaller than the ones of the technical data of the module (right), same behavior can be observed for voltage values with equal electric current values in both graphs. The temperature difference can be explained as the result of the thermal resistance variables considered for the construction of the model (hot, cold side, thermal paste and thermal capacity of heat sinks); according to X. Hao et. al. [11] the higher the thermal resistance value (K/W) on the hot side of the Peltier module, the smaller the temperature difference between the hot and cold side.

5. Conclusions

The developed model allows better knowing of the Peltier module's behavior under different conditions of ambient temperature and electric power supply; moreover, the simulations result in graphs with similar results to those provided in the technical data of the module used as reference.

Therefore, the model represents an accurate and immediate way to simulate the electrical and thermal behavior of a Peltier module under certain operating conditions and it could be used to test if the module meets the necessary conditions for a specific application. One of the advantages of the model is the possibility to perform simulations of Peltier systems without physically building the system, inferring a time and resources reduction option.

The model has no limitations and can be adapted to any type of module or manufacturer. It is only necessary to know the characteristic data of the module such as the Seebeck coefficient, the electrical resistance, the thermal resistance, and the resistance of the thermal grease used between the heat sink and the module. It is also necessary to know data such as thermal resistance and thermal capacity of the heat sinks that will be coupled to the hot and cold side of the module and the temperature of the environment where the Peltier system will operate.

References

[1] M. Chen, G. Jeffrey Snyder, Analytical and numerical parameter extraction for compact modeling of thermoelectric coolers, International Journal of Heat and Mass Transfer. **60** (2013) 689-699.
[2] S. Lineykin, S. Ben-Yaakov, Spice compatible equivalent circuit of the energy conversion processes in thermoelectric module, IEEE Israel Convention. **23** (2004) 346-349.
[3] D. Mitrani, J. Salazar, A. Turó, M. J. García, J. A. Chávez, Lumped and distributed parameter spice models of te devices considering temperature dependent material properties, Therminic, 2007, pp. 202–207.
[4] R. Ahiska, H. Mamur, A review: thermoelectric generators in renewable energy, Int J Renew Energy Res, **4**(1) 2014 29-36.
[5] H. L. Tsai, J. M. Lin, Model Building and Simulation of Thermoelectric Module Using Matlab/Simulink, Journal of Electronic Materials. (2010) 2105-2107.
[6] D. Zhao, G. Tan, A review of thermoelectric cooling: materials, modeling and applications, Appl Therm Eng. **66** (1-2) (2014) 15-24.
[7] G. Min, D. M. Rowe, Experimental evaluation of prototype thermoelectric domestic-refrigerators, Appl. Energy, 2006, pp. 133-152.
[8] H. S. Choi, S. Yun, K. I. Whang, Development of a temperature-controlled car-seat system utilizing thermoelectric device, Applied Thermal Engineering. **27**(17), (2007) 2841-2849.
[9] G. A. Mannella, V. La Carrubba, V. Brucato, Peltier cells as temperature control elements: Experimental characterization and modeling, Applied Thermal Engineering. **63** (2014) 234-245.
[10] K. Ogata, Ingeniería de Control Moderna, Prentice Hall, 2010.
[11] X. Hao, B. Peng, G. Xie, Y. Chen, Efficient on-chip hotspot removal combined solution of thermoelectric cooler and mini-channel heat sink, Applied Thermal Engineering. **100** (2016) 170-178.

Acknowledgements

This publication was supported by COFFA, Instituto Politécnico Nacional (IPN) and CCYTET-Tabasco, Mexico.

AUTHOR INDEX

Afolalu, S.A. 12
Cabral-Albuquerque, Elaine C. M. 40
Chijuo, C. 12
Cunha, Silvio 40
Emetere, M.E. 6, 12
Fah, TEE Kong 1
Fialho, Rosana L. 40
Flores-Vela, Abelardo I. 45
Hafsa, Sayera 24
Ikechukwu, M. Ahiara 6
Jerzy, SZOLOMICKI 29
Jimenez-Rodriguez, David J. 45
Kazmi, Zaheer Abbas 35
Leão, Mirtânia A. 40
Li, Wang 1
Majid, Mazlina Abdul 24
Marquez-Rocha, Facundo J. 45
Miranda Jr., Eval O. 40
Palma-Ramirez, Diana 45
Paredes-Rojas, Juan Carlos 45
Sanchez-Ramos, Carlos A. 45
Silva, Victor H. R. 40
Svetlana, Teplinskaya 18
Willie, Ahli K. D. 18
Yongfeng, Fang 1
Zhao, Hongtao 18

Institute of Physics
Dirac House, Temple Back
Bristol BS1 6BE UK

ISSN: 1757-8981
ISBN 978-1-7138-3171-6
9 781713 831716

CURRAN ASSOCIATES INC.
proceedings
.com